PHYSICS PROBLEMS:
Electricity, Magnetism, and Optics

PHYSICS PROBLEMS:
Electricity, Magnetism, and Optics

Robert L. Gray
University of Massachusetts

John Wiley & Sons, Inc.
New York • London • Sydney • Toronto

Library of Congress Cataloging in Publication Data

Gray, Robert L
Physics problems: electricity, magnetism, and optics.

(Wiley self-teaching guides)
1. Electricity–Problems, exercises, etc. 2. Magnetism–Problems, exercises, etc. 3. Optics–Problems, exercises, etc. I. Title.
QC532.G7 530 73-20445
ISBN 0-471-32411-6

Printed in the United States of America

74 75 10 9 8 7 6 5 4 3 2 1

To the Reader

Physics students often complain that they understand the physics but cannot solve problems. While their professors might think that impossible, the dilemma is real to the student. This book is an attempt to bridge that very real gap.

Physics Problems: Electricity, Magnetism, and Optics is designed to engage you in a discussion of physics problems as if you were in a small discussion class. The style is conversational and the format should help you sort out your own strengths and weaknesses as you actually work on problems.

Successful problem-solvers have at one time or another learned to do three things:

(1) Make a serious attempt at solving a problem and thereby benefit from successes and failures which occur.
(2) Obtain feedback from a teacher, a fellow student, or a book that aids in identifying and eliminating errors.
(3) Criticize their own work, requiring that solutions make sense and that they be consistent with some general conceptual framework.

This book is intended to give you a great deal of practice in doing the first two and some of the third. The format will give you feedback in the form of answers to questions and will allow you to bypass sections that you don't need. The book assumes that you are or have been enrolled in a physics course, so it does not attempt to develop concepts as fully as a textbook would. In each chapter, a Programmed Study Section reviews the necessary physical concepts, emphasizing the high points of particular topics, the meaning of symbols, graphical analysis, mathematical techniques, and diagrams used to present these concepts.

Physic Problems: Electricity, Magnetism, and Optics covers topics normally included in the second semester of a physics course. A separate, self-contained book, *Physics Problems: Mechanics and Heat,* covers topics from the first half of a year-long course in physics. Each book is independent, so they can be used separately or together.

How to Use This Book

To make this book work best for you, you should be familiar with its flexible format. Each of the chapters has four sections:

- Sample Problems and Objectives
- Programmed Study Section
- Programmed Solutions to Sample Problems
- Self-Test

The Sample Problems and Objectives give a preview of the content in each chapter. First try to work the sample problems and then compare your solutions with the answers on the next page. If you can solve all the problems and if the objectives are familiar, you can probably skip the chapter or skim it quickly. Be sure you read the objectives. Even if the problems seem easy, the discussion often covers techniques that will help you solve more difficult types of problems.

If either the problems or the objectives cover unfamiliar material or if your comprehension is shaky, turn to the Programmed Study Section which follows. This section reviews the concepts and techniques basic to solving the problems in the chapter. The discussion is presented in numbered frames. Each frame will present some information and ask you a question or give you a problem to solve. By checking your answer with the one given below the dashed line, you can assure that you understand each part of the discussion. If your answer is different from the one given, be sure you understand why before you go on to the next frame. An explanation is often provided as well as the answer. These explanations should be considered carefully if you have given an incorrect response.

Even if you can correctly answer the sample problems, it may be useful to go through the Programmed Study Section comparing your technique with that given. The combination of techniques may be more effective than either one individually. There is more to be learned than just the answer to a specific problem.

The third section presents a step-by-step solution to each of the sample problems. This section is also programmed so that you actually work out the problem yourself, checking your progress at each step. If you still have difficulty, you may wish to reread the Programmed Study Section.

Finally, the problems in the Self-Test will allow you to test your ability to solve problems like those discussed in the chapter. See if you are able to handle the problems, checking the answers which follow.

If you want to review a particular type of problem or a problem in a certain area of physics, look up the topic in the Index. Turn to the appropriate chapter and read the sample problems. When you find the type of problem you're looking for, turn to the solution of that sample problem.

You will be able to write your answer in the spaces provided in each frame. However, room has not been left for all calculations and lengthy solutions, so you may wish to keep some scratch paper handy. This book will help most if you actually try to answer the questions in each frame. Cover the printed answer and then compare your answer with the one given.

This book is not, of course, a physics textbook; rather, it assumes that you are either taking or have taken a physics course. So the Programmed Study Sections only highlight those techniques and concepts necessary to problem-solving in that chapter. But if you want to read fuller explanations on a particular topic, the Cross-Reference Chart on page xiii will help you locate the appropriate pages in some of the major physics texts.

Contents

REFERENCES FOR SELECTED TEXTBOOKS IN INTRODUCTORY PHYSICS*

Atkins, Kenneth R., *Physics,* 2nd edition (New York: John Wiley & Sons, Inc., 1970).

Bueche, Frederick, *Introduction to Physics for Scientists and Engineers* (New York: McGraw-Hill Book Company, 1969).

Freeman, Ira M., *Physics: Principles and Insights* (New York: McGraw-Hill Book Company, 1973).

Gamow, George, and Cleveland, John, *Physics, Foundations and Frontiers* (New Jersey: Prentice-Hall, Incorporated, 1969).

Halliday, D., and Resnick, R., *Fundamentals of Physics* (New York: John Wiley & Sons, Inc., 1970).

Marion, Jerry B., *Physics and the Physical Universe* (New York: John Wiley & Sons, Inc., 1971).

Miller, Franklin Jr., *College Physics* (New York: Harcourt Brace Jovanovich, Inc., 1972).

Sears, Francis W., and Zemansky, M. W., *University Physics,* 2 pts, 4th edition (Reading, Massachusetts: Addison-Wesley Publishing Company, Inc., 1970).

Weidner, Richard T., and Sells, Robert L., *Elementary Classical Physics,* 2nd revised edition, Vol. 1 and 2 (Boston: Allyn and Bacon, Inc., 1973).

*The mathematical level in *Physics Problems: Electricity, Magnetism, and Optics* is noncalculus; however, this book discusses many of the concepts and problems presented in textbooks which include calculus.

Chapter in this book	Atkins	Bueche	Freeman	Gamow & Cleveland	Halliday & Resnick	Marion	Miller	Sears & Zemansky	Weidner & Sells
1 Coulomb's Law	13	18	16	12	22	6	17	24	22
2 Electric Fields & Gauss's Law	14	18,19	16	12	23,24	8	18	25	23,24
3 Electric Potential	14	20	16	12	25	7,8	-	26	25
4 Magnetic Fields	16,17	24,25	18	14	29,30	9	21	30,31,32	29,30
5 Faraday's Law of Induction	-	26	19	14	31	-	21	33	31
6 Current Electricity	15	21,22	17	13	27,28	-	19,20	28,29	27,28
7 Electric Energy, Heat, and Power	15	21	17	13	27	-	19,20	28,29	27
8 Reflection and Refraction	-	31	14	15	36	-	24	38,39,40	36,37

PHYSICS PROBLEMS:

Electricity, Magnetism, and Optics

CHAPTER ONE
Coulomb's Law

If the sample problems and objectives below identify your weak points, go directly to the programmed study section on page 2. If not, try the problems and compare your answers with those that follow. If you can do all the problems easily and if you are familiar with the objectives, you may wish to skip all or part of this chapter. The programmed study section covers techniques and concepts basic to solving the sample problems and fulfilling the objectives in this chapter. A programmed, step-by-step solution of each sample problem begins on page 5. A self-test is included at the end of the chapter.

SAMPLE PROBLEMS AND OBJECTIVES

Problem 1

Four charges are arranged as shown. Find the electrostatic force exerted on the charge at the origin.

$a = 1.0$ m, $q = 2.0 \times 10^{-6}$ coul

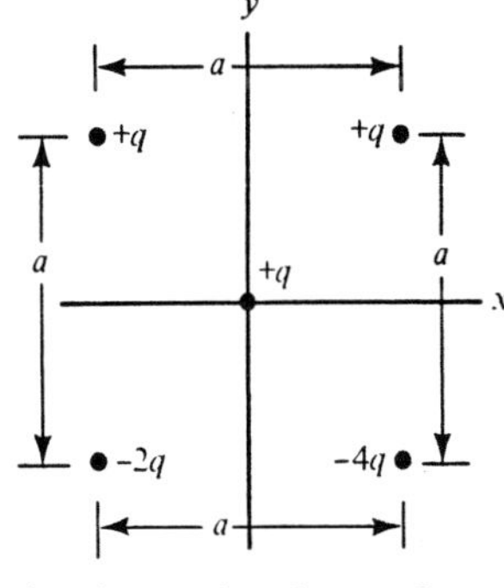

Objectives:
1. Reviewing the vector nature of electrostatic forces.
2. Reviewing right triangles.

Problem 2

Two positive charges separated by a distance of 0.1 m repel each other with a force of 18 nt. What is the charge on each if the total charge is 9.0×10^{-6} coul?

Objectives:
1. Solving problems algebraically prior to numerical substitution.
2. Solving quadratic equations.

Problem 3

The hydrogen atom can be pictured as an electron moving with constant speed in a circular orbit about a proton. The orbital radius is 5.3×10^{-11} m. This motion of the electron can be considered a current flow. What is the value of the electron current in coul/sec?

Objectives:

1. Using Coulomb's law in a "practical" case in which typical numerical values are discussed.
2. Reviewing the dynamics of circular motion.
3. Introducing the concept of electric current.

Answers to Sample Problems

See page 5 for programmed, step-by-step solutions to these problems.

Problem 1

$F_{\text{total}} = 4.21 \times 10^3$ nt at an angle of 14^0 to the right of the $-y$ axis.

Problem 2

$q_1 = 5 \times 10^{-6}$ coul
$q_2 = 4 \times 10^{-6}$ coul

Problem 3

$I = 1.0 \times 10^{-3}$ coul/sec

PROGRAMMED STUDY SECTION

1 Let us review the basic elements of Coulomb's law. Below are shown three different cases in which two point charges interact with each other. In each case indicate by a vector the Coulomb force acting on each charge.

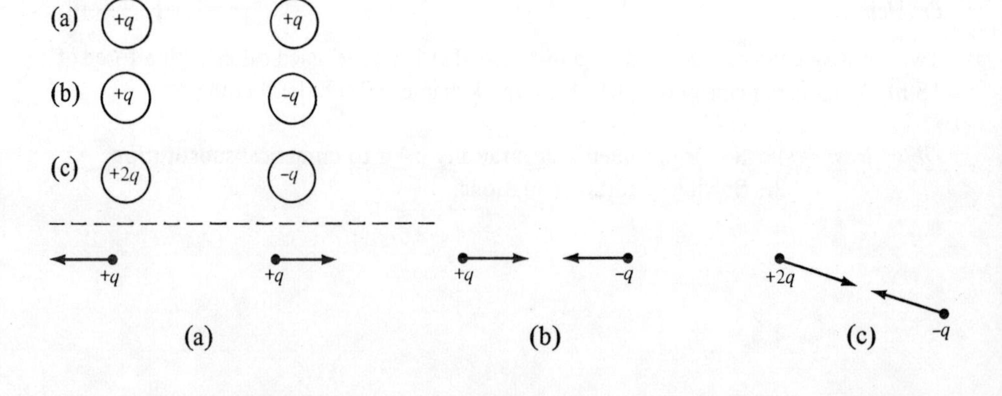

2 As in the examples of the first frame, like charges "repel" and unlike charges "attract." For the case of two charges interacting with each other the ________________ of the Coulomb force acting on each charge is the same. The ________________ of the Coulomb force on each is along the line joining the two charges with the direction determined by the types of charges involved.

---

magnitude; direction

3 For the configuration of charges shown, indicate by vectors the electrostatic forces exerted on q_1 by q_2 and by $-q_3$. Indicate only the forces on q_1. All charges have the same magnitude.

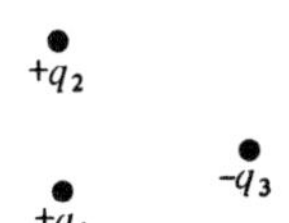

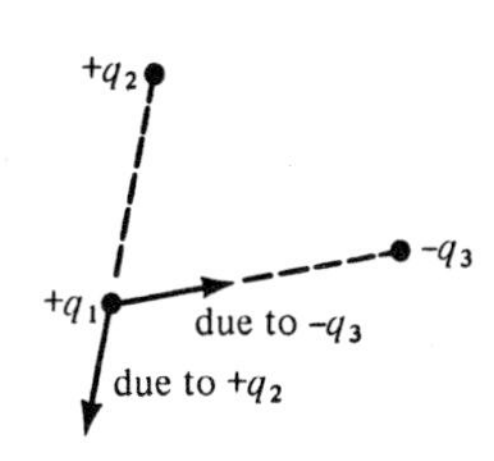

4 Here we have labeled as $\mathbf{F}_{12}$ the electrostatic forces acting on q_1 by its interaction with q_2. $\mathbf{F}_{13}$ is similarly the force on q_1 by q_3. The electrostatic forces $\mathbf{F}_{12}$ and $\mathbf{F}_{13}$ are vector forces acting on q_1. To completely specify these vectors we need to know the magnitude and direction of each.

Will the directions (senses) of the forces in the diagram be as shown even if the magnitudes of the charges are altered?

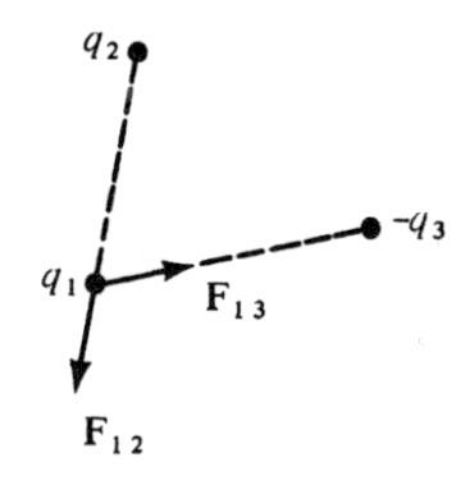

Yes. The directions of the forces are determined by the signs (like or unlike) of the paired charges and their configuration (i.e., the line joining the pair). The directions are not determined by the magnitudes of the charges.

5 Although we have not selected a reference direction (i.e., coordinate system), we do know the directions of the electrostatic forces between q_1 and q_2 as well as between q_1 and q_3. In your own words, what is the direction of the electrostatic force exerted on one charge by a second charge?

Along the line joining the charges, being either repulsive or attractive depending on whether the charges are of the same or opposite sign

6 The quantity to be determined now is the magnitude of the electrostatic force. Write the scalar equation which gives the force on a charge q_1 due to another charge q_2. Assume the charges to be separated by a distance r_{12}.

F_{12} = ____________

$$F_{12} = \frac{1}{4\pi\epsilon_0} \frac{q_1 q_2}{r_{12}^2}$$

This equation is known as Coulomb's law. Some specifics about the equation should be pointed out:

r_{12} is the distance from q_1 to q_2

$\frac{1}{4\pi\epsilon_0}$ is a constant equal to 9×10^9 nt-m²/coul² in the MKS system of units

7 Two point charges $q_1 = 4.0 \times 10^{-6}$ coul and $q_2 = -8.0 \times 10^{-6}$ coul are separated by 4 m as shown.

(a) What is the direction of the electrostatic force on q_1?
(b) What is the magnitude of the force on q_1? Obtain a numerical result.

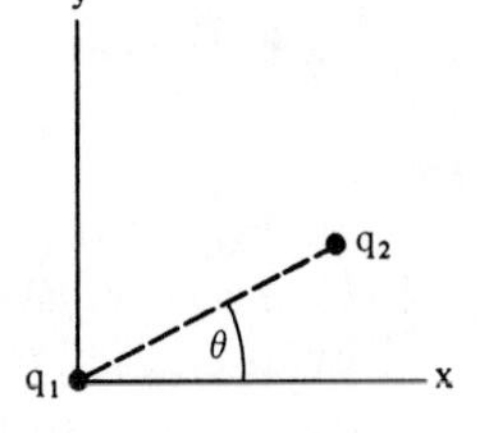

(a) Toward q_2 along the line connecting q_1 and q_2

(b) $F_{12} = \frac{1}{4\pi\epsilon_0} \frac{q_1 q_2}{r_{12}^2}$

$F_{12} = 18.0 \times 10^{-3}$ nt

8 Let θ in the problem be 30°. What are the x and y components of $\mathbf{F}_{12}$?

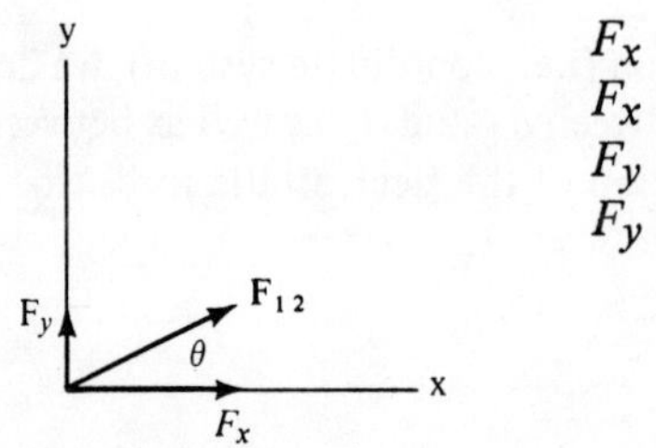

$F_x = F_{12} \cos\theta$
$F_x = 15.6 \times 10^{-3}$ nt
$F_y = F_{12} \sin\theta$
$F_y = 9.0 \times 10^{-3}$ nt

9 Describe fully the electrostatic force acting on q_2 due to q_1.

It is oppositely directed to $\mathbf{F}_{12}$ and of the same magnitude.

Note: When, as in frame 3, a charge is under the influence of more than one other charge, the total electrostatic force can be determined by finding the vector sum of all individual forces.

SOLUTIONS TO SAMPLE PROBLEMS

Problem 1

Four charges are arranged as shown. Find the electrostatic force exerted on the charge at the origin.

$a = 1.0$ m, $q = 2.0 \times 10^{-6}$ coul

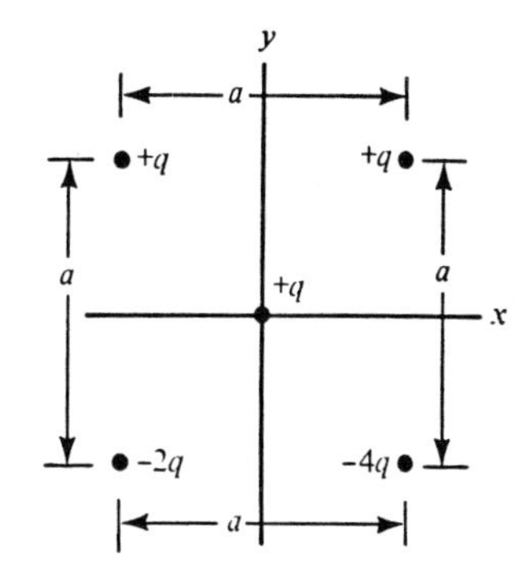

1.1 The approach to this problem invokes the principle of superposition. Superposition means that we can calculate the total force on the charge at the center by first pairing the central charge q with each of the four surrounding charges and then determining the vector sum of the four forces.

Draw vectors on the diagram representing the forces acting on $+q$ at the center due to its separate interactions with the four other charges. Label the force UL (due to upper left charge), LR (due to lower right charge), LL (due to lower left charge), or LR (due to lower right charge).

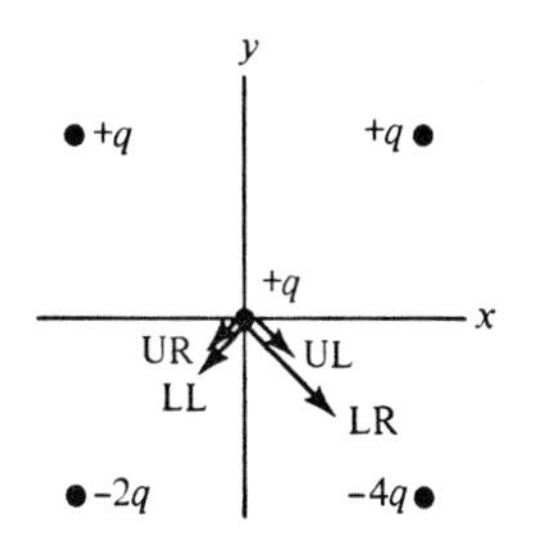

Coulomb's law can be used to calculate (for example) the force UL by ignoring the other three charges.

1.2 Now you know the directions of the four separate forces. Which forces have equal magnitudes? (Identify as UL, UR, LL, or LR.)

UL = UR

1.3 Prior to calculating the magnitude we need to do a bit of geometry. Here is a piece of the diagram from frame 1.1. Identify all the angles of this triangle as well as the length of the horizontal and vertical sides in terms of the given length a of the problem.

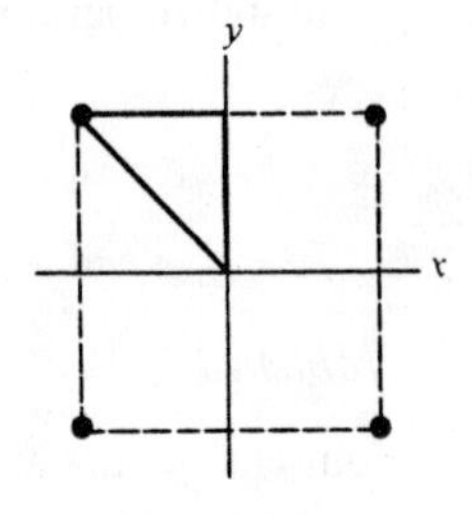

$\frac{a}{2}$ 45°   90° $\frac{a}{2}$ r 45°

1.4 From the theorem of Pythagoras, what is the hypotenuse r of the triangle? (You may choose to answer this question using trigonometry. Try it!)

$$r = \sqrt{\frac{a^2}{4} + \frac{a^2}{4}} \qquad \text{or} \qquad \frac{a}{2} = r \sin 45^\circ$$

$$r = \frac{a}{\sqrt{2}} \qquad\qquad \frac{a}{2} = (r)\left(\frac{1}{\sqrt{2}}\right)$$

$$r = \frac{a}{\sqrt{2}}$$

1.5 The separation between $+q$ at the center and each of the four charges is thus $a/\sqrt{2}$. From Coulomb's law then

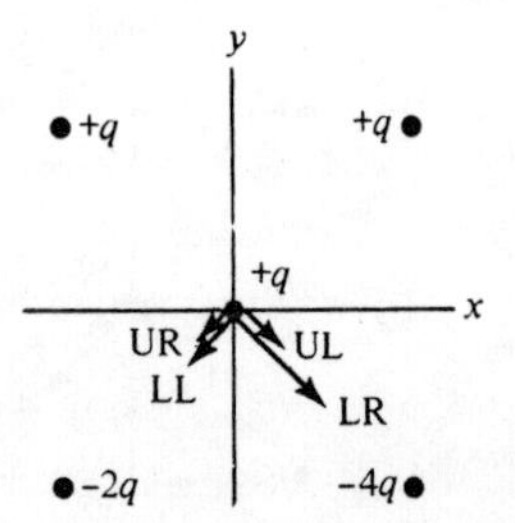

(a) F_{UL} = ________________

(b) F_{LL} = ________________

(c) F_{UR} = ________________

(d) F_{LR} = ________________

Having determined the directions of all the forces we need only use the magnitude of the charges to compute the forces from Coulomb's law.

(a) $F_{UL} = \frac{1}{4\pi\epsilon_0} \frac{q^2}{a^2/2}$

(b) $F_{LL} = \frac{1}{4\pi\epsilon_0} \frac{2q^2}{a^2/2}$

(c) same as (a)

(d) $F_{LR} = \frac{1}{4\pi\epsilon_0} \frac{4q^2}{a^2/2}$

1.6 Using the answers from the previous frame, perform the following additions.

(a) $F_{UL} + F_{LR}$ = ______________

(b) $F_{UR} + F_{LL}$ = ______________

(a) $\frac{1}{4\pi\epsilon_0} \left(\frac{q^2}{a^2/2} + \frac{4q^2}{a^2/2} \right) = \frac{1}{4\pi\epsilon_0} + \frac{10q^2}{a^2}$

(b) $\frac{1}{4\pi\epsilon_0} \frac{6q^2}{a^2}$

1.7 Why was it appropriate to add the magnitudes in the previous frame?

Because $\mathbf{F}_{UL}$ and $\mathbf{F}_{LR}$ were in the same direction (i.e., they are colinear); likewise, $\mathbf{F}_{UR}$ and $\mathbf{F}_{LL}$.

1.8 Here the diagram is somewhat simpler as we now have added those pairs of forces which are colinear. (This is not drawn to scale.)

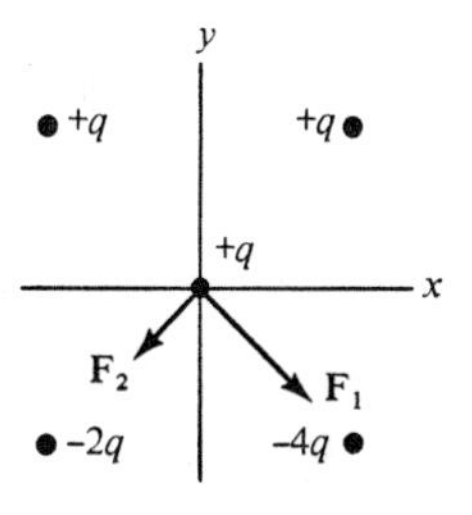

$$\mathbf{F}_1 = \mathbf{F}_{UL} + \mathbf{F}_{LR}$$
$$\mathbf{F}_2 = \mathbf{F}_{UR} + \mathbf{F}_{LL}$$

For $q = 2 \times 10^{-4}$ coul, $a = 1$ m, and $\frac{1}{4}\pi\epsilon_0 = 9 \times 10^9$ nt-m^2/coul^2

F_1 = ____________ nt

F_2 = ____________ nt

Go back to the answer of frame 1.5 and put the numbers in the answers. Combine vectors as appropriate.

$F_1 = 3.6 \times 10^3$ nt
$F_2 = 2.16 \times 10^3$ nt

For example

$$\mathbf{F}_1 = \mathbf{F}_{UL} + \mathbf{F}_{LR}$$

$$\mathbf{F}_1 = \frac{1}{4\pi\epsilon_0}\left(\frac{q^2}{a^2/2} + \frac{4q^2}{a^2/2}\right)$$ from the answer to frame 1.5

or

$$\mathbf{F}_1 = \frac{1}{4\pi\epsilon_0}\left(\frac{10\,q^2}{a^2}\right)$$

$$\mathbf{F}_1 = (9 \times 10^9 \text{ nt-m}^2/\text{coul}^2)\left(\frac{10 \times 4 \times 10^{-8} \text{ coul}^2}{1 \text{ m}^2}\right)$$

$$\mathbf{F}_1 = 3.6 \times 10^3 \text{ nt}$$

1.9 Vectors as now drawn are roughly to scale. Diagrammatically resolve $\mathbf{F}_1$ and $\mathbf{F}_2$ into their respective x and y components. We need to do this in order to find the final resultant force.

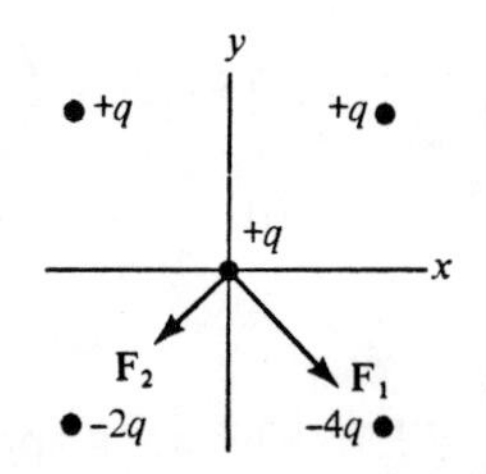

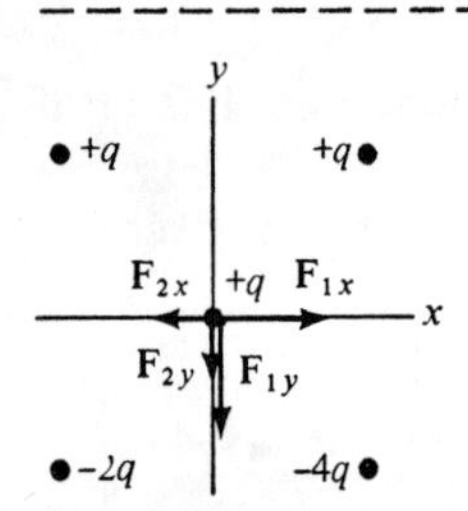

1.10 What is the angle which $\mathbf{F}_1$ and $\mathbf{F}_2$ make with the x axis?

From the geometry of the problem the angle is 45° in each case.

1.11 Looking at the answer of frame 1.9, choose the appropriate trig function for the following.

(a) $F_{1x} = F_1$ (sin or cos) 45°
(b) $F_{1y} = F_1$ (sin or cos) 45°

(a) $\cos 45°$; (b) $\sin 45°$

Note: $\cos 45° = \sin 45° = 1/\sqrt{2}$ or $\sqrt{2}/2$

1.12 Using the standard sign conventions and the answers of frame 1.11 calculate the following.

(a) $F_{1x} =$ ____________________

(b) $F_{1y} =$ ____________________

(c) $F_{2x} =$ ____________________

(d) $F_{2y} =$ ____________________

(a) $F_{1x} = F_1 \cos 45° = (3.6 \times 10^3 \text{ nt}) \left(\frac{\sqrt{2}}{2}\right) = 2.55 \times 10^3$ nt

(b) $F_{1y} = -2.55 \times 10^3$ nt

(c) $F_{2x} = -1.53 \times 10^3$ nt

(d) $F_{2y} = -1.53 \times 10^3$ nt

1.13 Now we are almost finished.

(a) $\Sigma F_x =$ ____________________

(b) $\Sigma F_y =$ ____________________

This can be done algebraically, of course. That is the utility of resolving forces into components.

(a) 1.02×10^3 nt

(b) -4.08×10^3 nt

1.14 Draw vectors representing the answers to the last frame and show their resultant diagrammatically. Use an approximate scale.

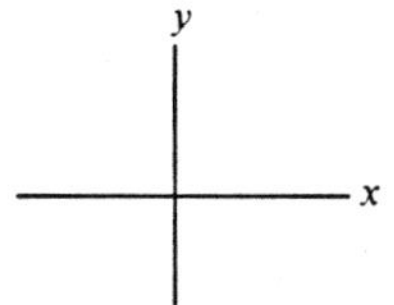

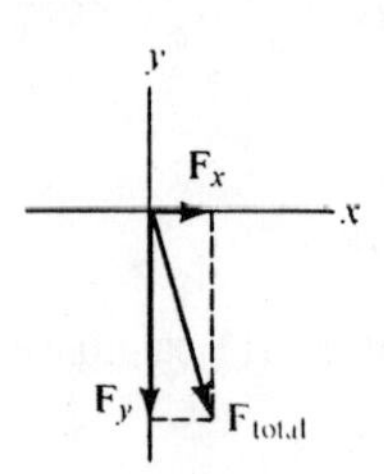

1.15 Finally, what is

(a) $\mathbf{F}_{total}$ in magnitude?
(b) Direction of $\mathbf{F}_{total}$ with respect to the $-y$ axis?

(a) $F_{total} = \sqrt{F_x^{\,2} + F_y^{\,2}} = 4.21 \times 10^3$ nt

(b)

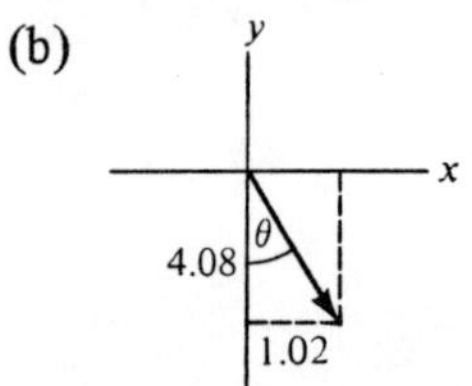

The total electrostatic force makes an angle θ to the right of the $-y$ axis. This angle is given by $\tan\theta = 1.02/4.08 = 0.25$. From tables we find that $\theta = 14°$.

Problem 2

Two positive charges separated by a distance of 0.1 m repel each other with a force of 18 nt. What is the charge on each if the total charge is 9.0×10^{-6} coul?

2.1 The two charges are separated by a distance r. If we call the lefthand charge q and the total charge Q, what is the righthand charge?

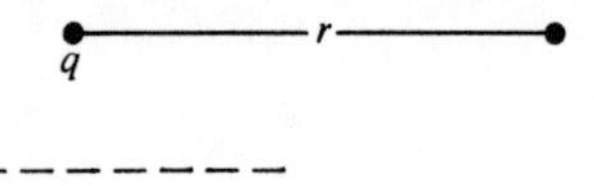

$(Q - q)$

If we were to call the charges q_1 and q_2 we would have two unknowns. Describing them as q and $(Q - q)$ results in only one unknown. Q is given in the problem.

2.2 Write Coulomb's law for the magnitude of the electrical force on the lefthand charge.

$$F_{\text{on } q} = \frac{1}{4\pi\epsilon_0} \frac{q(Q-q)}{r^2}$$

2.3 We know everything in the equation in the previous answer except q. The algebra is a little less cumbersome is we use $k = 1/4\pi\epsilon_0$. Later we will let $k = 9 \times 10^9$ nt-m^2/coul^2. The equation is thus

$$F = k \left[\frac{q(Q-q)}{r^2}\right]$$

We see in the numerator of the righthand side that we have a term in q and a term in q^2. Rewrite the equation in the form

$$aq^2 + bq + c = 0$$

where a, b, and c are constants.

$$q^2 - Qq + \frac{Fr^2}{k} = 0$$

2.4 Putting numbers into the equation (ignoring units for simplicity) gives

$$q^2 - (9 \times 10^{-6})\,q + \frac{18 \times 0.01}{9 \times 10^9} = 0$$

Use the quadratic formula to solve for q.

$q = 5 \times 10^{-6}$ coul

$q = 4 \times 10^{-6}$ coul

$$q = \frac{-b \pm \sqrt{b^2 - 4ac}}{2a}$$

$$q = \frac{(9 \times 10^{-6}) \pm (1 \times 10^{-6})}{2}$$

2.5 Note that there are two positive roots of the equation. Interpret these in terms of the original choice of the unknown q.

One could have called the righthand charge q, in which case the same answer would have resulted. The second root is the charge on the opposite side.

Problem 3

The hydrogen atom can be pictured as an electron moving with constant speed in a circular orbit about a proton. The orbital radius is 5.3×10^{-11} m. This motion of the electron can be considered a current flow. What is the value of the electron current in coul/sec?

3.1 The electron in the hydrogen atom moves about the proton as shown. This is an example of uniform circular motion. The centripetal force on the electron results from the electrical interaction.

Draw a vector to represent the force on the electron due to the proton.

- - - - - - - - - - - - - - - - - - - -

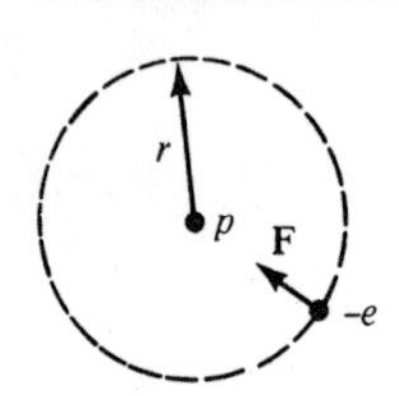

The force points toward the center (i.e., towards the proton).

3.2 In terms of F, m_e, v, and r, write Newton's second law for the motion of the electron.

- - - - - - - - - - - - - - - - - - - -

$$F = \left(m_e \frac{v^2}{r}\right)$$

3.3 The magnitude of F can be written in terms of the magnitude of the charges and their separation as the force is due to the electrostatic force between two charges.

$F =$ ______________________

- - - - - - - - - - - - - - - - - - - -

$$F = \frac{1}{4\pi\epsilon_0} \frac{e^2}{r^2} = m_e \frac{v^2}{r}$$

Both proton and electron have the same charge magnitude represented by e.

3.4 The Coulomb force

$$F = \frac{1}{4\pi\epsilon_0} \frac{e^2}{r^2}$$

for the electron is due to the two charges. If each charge has a magnitude 1.6×10^{-19} coul, what is the numerical value of the force on the electron?

- - - - - - - - - - - - - - - - - - - -

$F = 8.2 \times 10^{-18}$ nt

3.5 The number above gives some idea of the forces involved between protons and electrons at the atomic level. Using this number, determine the orbital speed of the electron. Use $m_e = 9.1 \times 10^{-31}$ kg. Obtain a number for the velocity.

$v = 2.2 \times 10^6$ m/sec

1.0 m/sec corresponds to about 2 mi/hr so it is clear that the electron is moving along nicely—something like 4,400,000 mi/hr!

3.6 If you could stand at one point along the orbit you would see the electron whiz by every so many seconds. How much charge would pass you during each revolution?

1.6×10^{-19} coul (the charge on the electron)

3.7 Since we know the orbital distance ($2\pi r$) and the orbital speed (v), we can calculate the time for one revolution. Obtain the period of the electron in seconds.

$T = \frac{2\pi r}{v} = 15.1 \times 10^{-17}$ sec

3.8 Electric current is defined as the total charge passing a given point per second. For the electron 1.6×10^{-19} coul passes a point on the orbit every 15.1×10^{-17} sec. What is the equivalent electric current in coul/sec?

$I \cong 1 \times 10^{-3}$ coul/sec

A current of 1.0 coul/sec is called an ampere. In the hydrogen atom we have a current of approximately a milliampere.

SELF–TEST

1 As shown to the right two pith balls of equal mass are suspended by silk threads of length ℓ. Each pith ball has the same charge. Show that the distance d is given by

$$d = \left(\frac{q^2 \ell}{2\pi\epsilon_0 mg}\right)^{\frac{1}{3}}$$

if the assumption is made that $\sin\theta = \tan\theta$ for θ not too large.

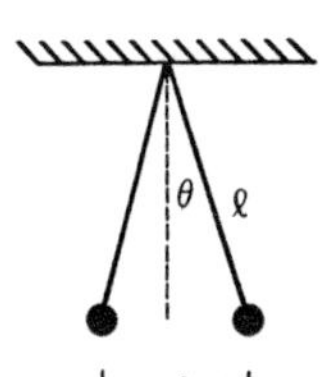

2 Three point charges are located at the corners of a right triangle as shown. Determine the magnitude and direction of the force on q_3 to each of the other charges.

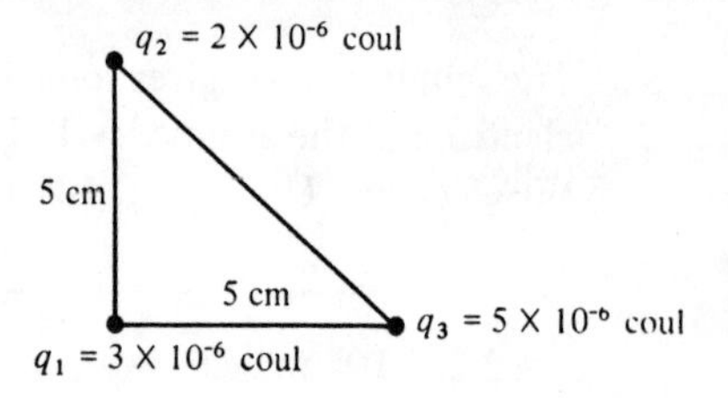

3 Calculate the magnitude of the repulsive force in pounds between two protons in a nucleus in which the separation is 4×10^{-15} m. Use the data below.

$$q = 1.6 \times 10^{-19} \text{ coul}$$
$$1 \text{ lb} = 4.4 \text{ nt}$$

Answers to Self-Test

1 For one of the charges at equilibrium (i.e., $\Sigma \mathbf{F} = 0$),

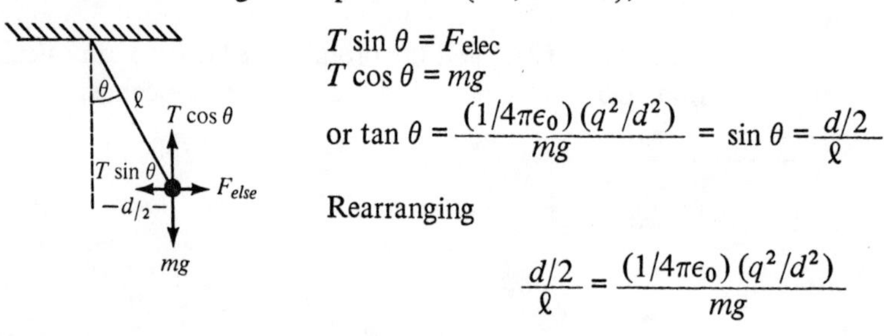

$$T\sin\theta = F_{\text{elec}}$$
$$T\cos\theta = mg$$

$$\text{or } \tan\theta = \frac{(1/4\pi\epsilon_0)(q^2/d^2)}{mg} = \sin\theta = \frac{d/2}{\ell}$$

Rearranging

$$\frac{d/2}{\ell} = \frac{(1/4\pi\epsilon_0)(q^2/d^2)}{mg}$$

yields the answer given.

2 54 nt due to q_1 acting along the line between q_1 and q_3 and pointing away from q_1; 18 nt due to q_2 acting along the line between q_1 and q_3 and pointing away from q_1

3 $F = 3.2$ lb, a large force for such small objects

CHAPTER TWO
Electric Fields and Gauss's Law

If the sample problems and objectives below identify your weak points, go directly to the programmed study section on page 17. If not, try the problems and compare your answers with those that follow. If you can do all the problems easily and if you are familiar with the objectives, you may wish to skip all or part of this chapter. The programmed study section covers techniques and concepts basic to solving the sample problems and fulfilling the objectives in this chapter. A programmed, step-by-step solution of each sample problem begins on page 25. A self-test is included at the end of the chapter.

SAMPLE PROBLEMS AND OBJECTIVES

Problem 1

Two charges q_1 and q_2 are arranged as shown. Calculate the electric field at point P.

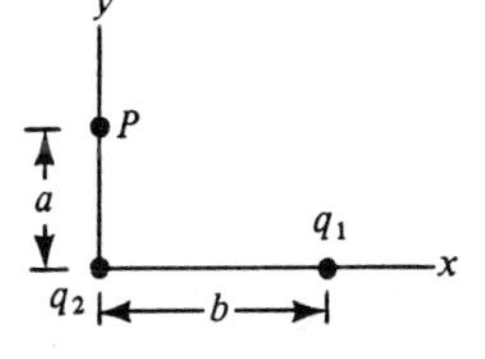

$q_1 = 2.0 \times 10^{-8}$ coul
$q_2 = 0.3 \times 10^{-8}$ coul
$a = 0.5$ m
$b = 1.0$ m

Objectives:
1. Calculating electric field of a distribution of point charges.
2. Reviewing vector addition.

Problem 2

A proton is accelerated from rest in a uniform electric field of strength 2.5×10^5 nt/coul. If the proton is accelerated over a distance of 0.4 m, what is the final kinetic energy and speed of the electron?

Objectives:
1. Discussing the dynamics and kinematics of charged particle motion in electric fields.
2. Numerical example of phenomena associated with particle accelerators.

Problem 3

An uncharged thin spherical metallic shell of radius $r = 5$ cm has a point charge $q = 3 \times 10^{-6}$ coul at the center of the shell. Use Gauss's law to determine the electric field inside and outside the metallic sphere. Determine numerically the field at a point 10 cm from the central charge.

Objectives:
1. Discussing Gauss's law.
2. Discussing induced charge.
3. Discussing electric fields in conductors.

Problem 4

Consider the cubical Gaussian surface shown in the diagram. The electric field in the coordinate system is

$$E_x = bx^{\frac{1}{2}}, E_y = 0, E_z = 0$$

Given that the cube measures 0.1 m on a side and $b = 800$ nt/coul-m$^{\frac{1}{2}}$, how much charge is inside the cube?

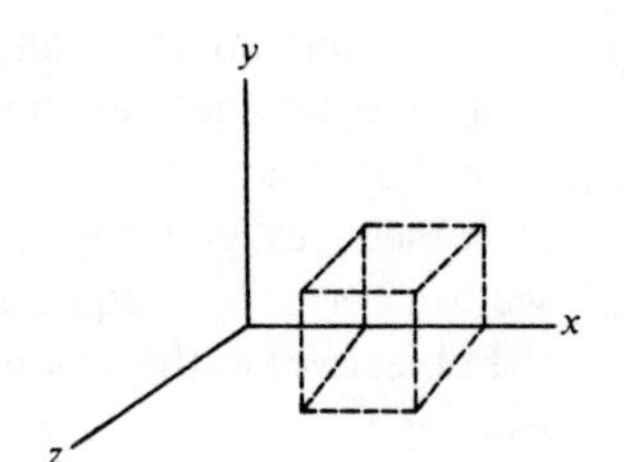

Objectives:
1. Reviewing the flux of an electric field.
2. Determining the charge using Gauss's law.

Answers to Sample Problems

See page 25 for programmed, step-by-step solutions to these problems.

Problem 1

$E_x = -128$ nt/coul
$E_y = 172$ nt/coul

Problem 2

(a) 1.6×10^{-14} joule
(b) 4.4×10^{6} m/sec

Problem 3

$E = 27 \times 10^{5}$ nt/coul, directed radially outward

Problem 4

$q = 9.7 \times 10^{-12}$ coul

PROGRAMMED STUDY SECTION

1 In this chapter you will be doing exercises similar to those of the previous chapter. The major difference is that now we will think in terms of the concept of the *electric field.* The first few frames will be used to develop this concept for the case of a point charge.

(a) What is the magnitude of the Coulomb force acting on q_2 as shown to the right?
(b) What is the direction of the force on q_2? (Show this by an appropriate vector.)

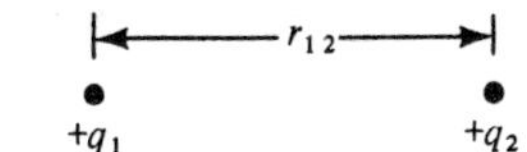

(a) $F_{21} = \dfrac{1}{4\pi\epsilon_0} \dfrac{q_1 q_2}{r_{12}^2}$

The subscript 21 means the force on q_2 by its interaction with q_1.

(b) r_{12}, $+q_1$, $+q_2$, $\mathbf{F}_{21}$

2 If we increase the magnitude of the charge q_2, the force $\mathbf{F}_{21}$ on q_2 will increase. We can rewrite $\mathbf{F}_{21}$ as

$$\frac{F_{21}}{q_2} = \frac{1}{4\pi\epsilon_0} \frac{q_1}{r_{12}^2}$$

The left side of this equation is the force per unit charge of q_2. For given r_{12} and q_1 the right side of this equation is a constant. Under these conditions, as q_2 increases F_{21} (increases/decreases). Hint: Remember that the spatial configuration remains fixed.

Increases (If the right side of the equation is a constant, then the left side must remain a constant.)

3 Looking again at the equation

$$\frac{F_{21}}{q_2} = \frac{1}{4\pi\epsilon_0} \frac{q_1}{r_{12}^2}$$

we introduce the concept of an electric field to interpret the left side of the equation. We define the term F_{21}/q_2 to be

(a) the electric field strength due to (q_1/q_2)
(b) at the distance r_{12} as measured from (q_1/q_2)

(a) q_1; (b) q_1 (We are talking about the electric field due to q_1 at a distance r_{12} away from q_1. We think of q_2 as a charge in the electric field of q_1.)

4 This is the new concept in this chapter. We associate with the space around a point charge q the *vector field* **E**. The magnitude of this field due to the charge q at a distance r away from q is

$E =$ ______________________

$$E = \frac{1}{4\pi\epsilon_0} \frac{q}{r^2}$$

This follows from Coulomb's law.

5 Since the electric field is a vector field we need to specify its direction as well as its magnitude. To the right we represent the electric field in the space around a positive charge q. From your understanding of Coulomb's law, what would be the direction of the force exerted on the positive charge q_0 which is shown in the electric field of q? (Show this by a vector representing the force **F** on q_0.)

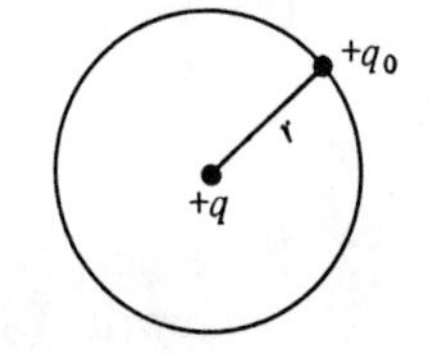

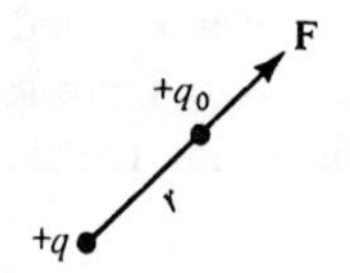

The vector is directed away from q since the charges are both positive. For any position of q_0, the force **F** would be directed radially away from q.

6 The direction of **F** on the positive charge q_0 in the previous frame is to be taken as the direction of **E**. The charge q_0 is called a *test charge.*

Now completely specify the electric field of q at the position of q_0 as shown in frame 5.

(a) $E =$ ______________________

(b) The direction of **E** is ______________________.

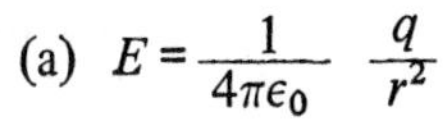

(a) $E = \frac{1}{4\pi\epsilon_0} \frac{q}{r^2}$

(b) directed radially away from q

For the point charge we represent electric field lines as shown.

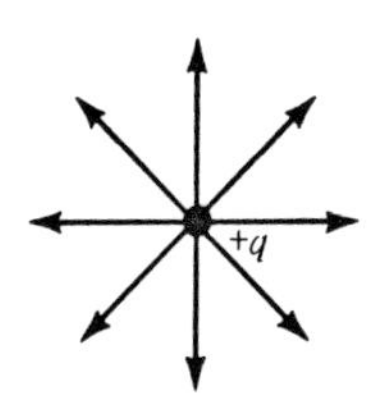

7 We can summarize the discussion so far with the equation

$$\mathbf{E} = \frac{\mathbf{F}}{q_0}$$

The idea here is that the test charge q_0 can be used to investigate a region of space for the existence of electric fields. If a force **F** (other than gravitational, magnetic, etc.) is observed to act on q_0 then both the magnitude and direction of **E** can be determined. Answer the following referring to the equation above.

(a) **E** and **F** are in the same direction. (True/False)
(b) **E** is the field due to q_0. (True/False)
(c) **E** and **F** have the same magnitude. (True/False)
(d) **F** is the force acting on q_0. (True/False)

(a) True, provided q_0 is positive.
(b) False. **E** is the field due to charges *other* than q_0.
(c) Faise. Look at the defining equation.
(d) True.

8 From the defining equation for **E** as given in frame 7, the units of **E** are ______________ ______________________________ in the MKS system.

---

newtons/coulombs

$\mathbf{E} = \mathbf{F}/q_0$
The units of **F** are nt. The units of q_0 are coul. The units of **E** are therefore nt/coul.

9 In problems involving a distribution of point charges we can find the total electric field at some point P by using the following equation:

$$\mathbf{E}_{\text{at } P} = \sum_{i=1}^{\eta} \mathbf{E}_i$$

In your own words, what does this equation mean?

The total electric field **E** is the *vector sum* of the electric field due to *each individual* charge. For example, $\mathbf{F}_1$ at P is determined as though q_1 were the only charge present.

10 For example one might have three point charges as shown in the diagram below.

(a) Draw vectors representing the direction of the individual electric fields due to q_1, q_2, and $-q_3$ at the point P. Label them $\mathbf{E}_1$, $\mathbf{E}_2$, and $\mathbf{E}_3$.

(b) E_1 = ______________

E_2 = ______________

E_3 = ______________

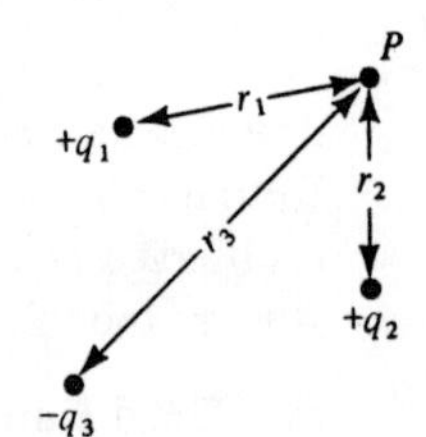

(a)

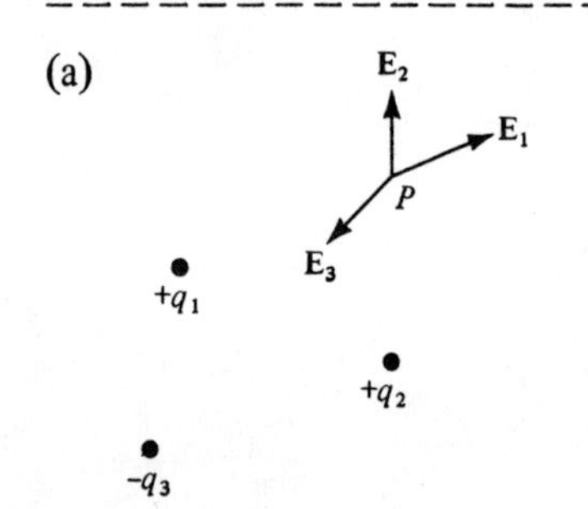

(b) $E_1 = \dfrac{1}{4\pi\epsilon_0}\,\dfrac{q_1}{r_1^{\,2}}$; $E_2 = \dfrac{1}{4\pi\epsilon_0}\,\dfrac{q_2}{r_2^{\,2}}$; $E_3 = \dfrac{1}{4\pi\epsilon_0}\,\dfrac{q_3}{r_3^{\,2}}$

The total electric field would be the vector sum $\mathbf{E}_1 + \mathbf{E}_2 + \mathbf{E}_3$.

11 Many problems involving uniform charge distributions are best handled using Gauss's law. But first we need some background material. The next few frames deal with things like flux, enclosed charges, surface elements, and normal components. A few general questions may help to untangle these ideas.

The balloon to the right is a closed surface. An element of area of the balloon is shown and is labeled $\Delta\mathbf{S}$. Draw on the sketch a vector representing this element of area.

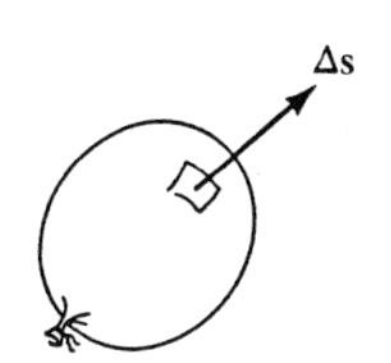

By convention the vector $\Delta\mathbf{S}$ is *perpendicular* to the surface element and points *outward.*

12 In using Gauss's law we want to calculate the flux of the electric field through a closed surface. Let us first consider the general notion of flux. To make things even simpler we won't consider the entire balloon. Rather we will take only the area element $\Delta\mathbf{S}$.

To the right we represent a constant electric field by evenly spaced field lines. What is the flux of **E** through the area element $\Delta\mathbf{S}$ (i.e., how many lines of **E** cross the surface in the direction of $\Delta\mathbf{S}$)?

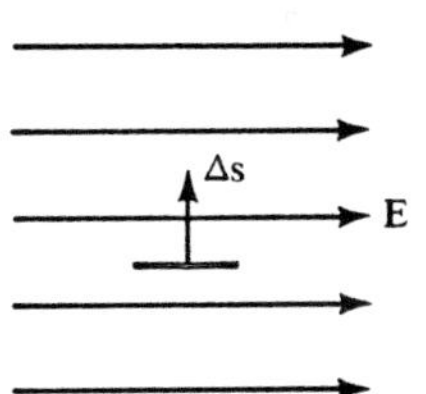

Zero (We say that the flux of **E** through the surface element $\Delta\mathbf{S}$ is zero.)

13 If we rotate the element of area 90° the flux is no longer zero. In this case what two things determine the flux of **E** through the area $\Delta\mathbf{S}$?

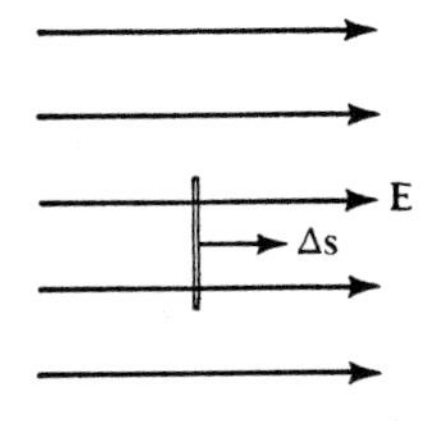

---

1. The magnitude of **E** (how many lines of **E** in the example)
2. The magnitude of the area element $\Delta\mathbf{S}$ (the size of the area)

14 Here we have the intermediate case. Now we must take into account the relative orientations of **E** and **ΔS**. The shorthand way to write this is:

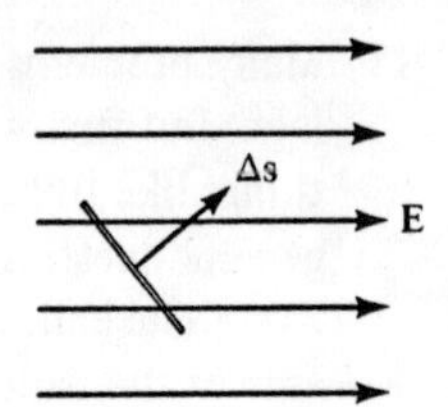

flux of **E** = ________________

E·ΔS (the so-called dot product)

In words this means component of **E** in the direction of **ΔS** times **ΔS**. This is similar to the idea of force times displacement when calculating work in mechanics.

E	×	ΔS	×	$\cos\theta$	=	flux of **E** through **ΔS**
strength of **E**		magnitude of the area element **ΔS**		relative orientation		

15 Gauss's law can be written:

$$\Sigma \mathbf{E}\cdot\Delta\mathbf{S} = \frac{q}{\epsilon_0}$$

(a) The lefthand side means the total flux of the electric field through a ________ surface.

(b) The meaning of q is ________________

(a) closed; (b) the net charge enclosed by the surface.

16 To the right is shown a charged plate which gives rise to a constant field **E** as represented by the field lines. Consider the closed surface of the imaginary balloon. There are charges only on the plate.

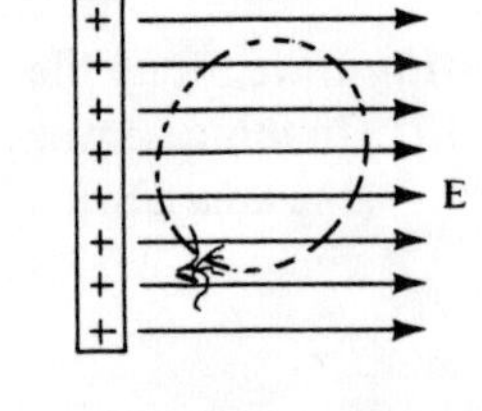

(a) If you had a device to detect electric fields, would you detect a field if you were "inside" the imaginary balloon? (Yes/No)

(b) The q on the righthand side of the Gauss's law is the net charge enclosed by the imaginary balloon surface. What *net* charge is *enclosed* by the surface?

(a) Yes (There are field lines inside the surface.)

(b) Zero (There are charges only on the plate.)

17 Look at Gauss's law with respect to the previous frame.

$$\Sigma \mathbf{E} \cdot \Delta \mathbf{S} = \frac{q}{\epsilon_0}$$

(a) The left side is the ______________ of **E** through the closed surface.
(b) If q is zero, what is the flux of **E**?

(a) flux (The symbol Φ is frequently used for the flux of **E**: $\Phi_E = \Sigma \mathbf{E} \cdot \Delta \mathbf{S}$.)
(b) Zero (Gauss's law is an equation relating flux and **E** to charge enclosed.)

18 The crucial point illustrated by the last two frames is that the flux of **E** may be zero, but this does not necessarily mean that **E** is zero. In the diagram of frame 16 the flux of **E** through the closed surface is zero, but **E** is not zero inside the surface. The distinction is so important that we must look at our example a little closer.

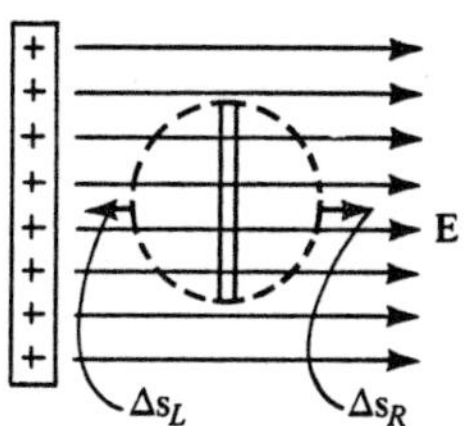

To the right we have divided the Gaussian surface into a lefthand part and a righthand part. Consider the flux contributions for the small area elements ΔS_L and ΔS_R.

(a) $\mathbf{E} \cdot \Delta \mathbf{S}_L$ = ______________

(b) $\mathbf{E} \cdot \Delta \mathbf{S}_R$ = ______________

(c) Considering that E and ΔS are the same in both cases, in what way do these two terms differ?

(a) $\mathbf{E} \cdot \Delta \mathbf{S}_L = -E\Delta S$ because the cosine of the angle is -1 ($\theta = 180°$).
(b) $\mathbf{E} \cdot \Delta \mathbf{S}_R = +E\Delta S$ because the cosine of the angle is +1 ($\theta = 0°$).
(c) The two terms thus differ only in sign, *but the difference is important.*

19 Gauss's law requires that we add all flux contributions of the type discussed in the preceding frame. Can you now say why the total flux through our balloon Gaussian surface will be zero?

The left half will have negative terms which when added to the positive terms for the right half will result in a total flux of zero. We have, of course, only looked at two specific terms in the sum.

20 Consider now a situation in which the electric flux through a closed surface is not zero.

At the right is shown a cubical closed surface. The electric field is perpendicular to the right and left side of the cube. Is the electric field the same on both the right and left faces of the cube? (For the moment, do not concern yourself about the situation within the cube.)

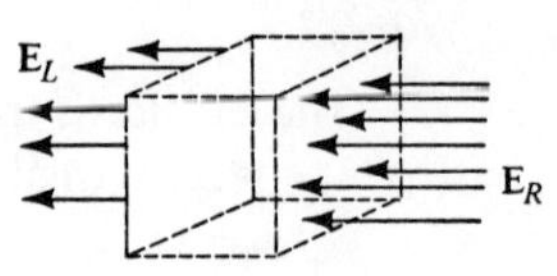

- - - - - - - - - - - - - - - - - -

No (In representing electric fields by field lines, a greater density of lines means a stronger field. The field is stronger at the right face than at the left.)

21 Draw on the sketch above two vectors to represent elements of area $\Delta\mathbf{S}$ on both the left and right faces of the cubical.

- - - - - - - - - - - - - - - - - -

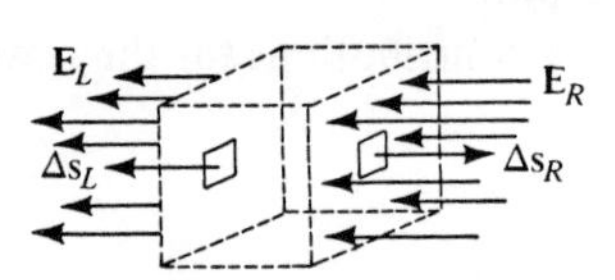

Note that both vectors $\Delta\mathbf{S}$ point "out" from the closed surface. This is the standard convention.

22 The algebraic sign of the electric flux through the right face is ______________.

The sign of the flux through the left face is ______________.

- - - - - - - - - - - - - - - - - -

negative; positive

Right face: $\mathbf{E}\cdot\Delta\mathbf{S}_R = E\Delta S_R \cos 180°$
Left face: $\mathbf{E}\cdot\Delta\mathbf{S}_L = E\Delta S_L \cos 0°$

23 The magnitude of the electric flux through the right side is (larger/smaller) than the electric flux through the left side.

- - - - - - - - - - - - - - - - - -

larger (because E is larger on the right side)

24 Thus, we see that the negative flux through the right side is larger in magnitude than the positive flux through the left side. The flux through the entire cube is

$$\Phi_E = \Sigma_{\text{cube}}\mathbf{E}\cdot\Delta\mathbf{S} = \underset{\text{side}}{_{\text{Left}}}\Sigma\mathbf{E}\cdot\Delta\mathbf{S} + \underset{\text{side}}{_{\text{Right}}}\Sigma\mathbf{E}\cdot\Delta\mathbf{S}$$

Will the answer be positive, negative, or zero? Note that all other sides of the closed surface are left out because their contribution to the flux is zero (i.e., no electric field lines pass through any of the other four sides).

Negative

25 Gauss's law states that the electric flux through a closed surface is proportional to the charge enclosed by that surface.

(a) Is there a net charge inside the cubical surface?
(b) If so, can you tell what the sign of the net charge is?

(a) Yes, there must be. (If there were no charge, the electric flux would have been zero.)
(b) Since the flux is negative, the net charge is also negative. There may be some positive as well as negative charges, but the *net* charge is negative.

SOLUTIONS TO SAMPLE PROBLEMS

Problem 1

Two charges q_1 and q_2 are arranged as shown. Calculate the electric field at point *P*.

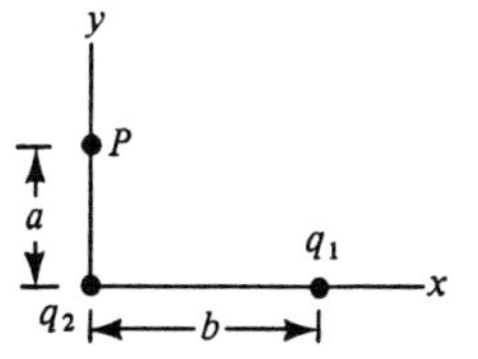

$q_1 = 2.0 \times 10^{-8}$ coul
$q^2 = 0.3 \times 10^{-8}$ coul
$a = 0.5$ m
$b = 1.0$ m

1.1 Presumably by now you have had many opportunities to find vector sums. You know that this involves coordinate systems, vector components, geometry, and trigonometry. All these techniques are still appropriate to finding the vector sum of electric fields.

Two positive point charges q_1 and q_2 are located in a coordinate system as shown. We seek the electric field at the point *P*. We know that the magnitude of the electric field a distance r away from a point charge is

$$E = \frac{1}{4\pi\epsilon_0} \frac{q}{r^2}$$

(a) What is r^2 for q_2 in this problem?
(b) What is r^2 for q_1 in this problem?

(a) a^2; (b) $a^2 + b^2$

Note that the answer in (b) is *not* $\sqrt{a^2 + b^2}$

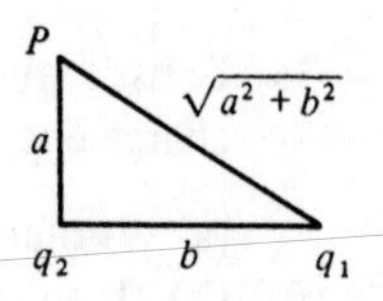

1.2 Subscript 1 means the electric field due to q_1. What are the expressions for the electric field magnitudes at the point P?

(a) E_1 = ____________________

(b) E_2 = ____________________

(a) $E_1 = \dfrac{1}{4\pi\epsilon_0}\dfrac{q_1}{a^2 + b^2}$; (b) $E_2 = \dfrac{1}{4\pi\epsilon_0}\dfrac{q_2}{a^2}$

1.3 For the case of $q_1 = 2.0 \times 10^{-8}$ coul, $q_2 = 0.3 \times 10^{-8}$ coul, $a = 0.5$ m, and $b =$ 1.0 m, find the following. ($1/4\pi\epsilon_0 = 9.0 \times 10^9$ nt-m^2/coul^2. Obtain numerical answers.)

(a) E_1 = ____________________

(b) E_2 = ____________________

(a) E_1 = 144 nt/coul; (b) E_2 = 108 nt/coul

1.4 Draw vectors showing the directions of $\mathbf{E}_1$ and $\mathbf{E}_2$ at P.

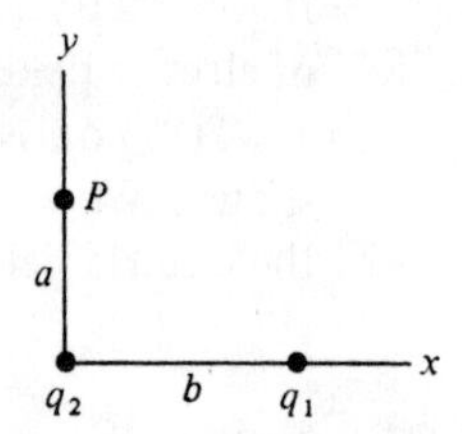

1.5 The problem is now to find the vector sum of $\mathbf{E}_1$ and $\mathbf{E}_2$. This is most conveniently done by resolving $\mathbf{E}_1$ into x and y components.

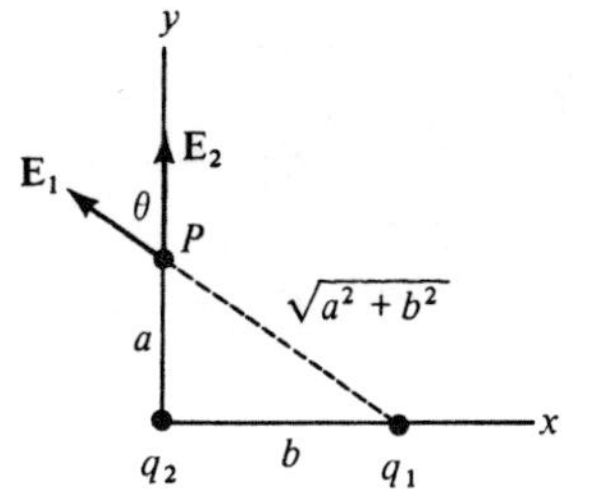

(a) In terms of the angle θ draw vectors on the diagram representing $\mathbf{E}_{1x}$ and $\mathbf{E}_{1y}$.

(b) $\mathbf{E}_{1x}$ = ____________________

(c) $\mathbf{E}_{1y}$ = ____________________

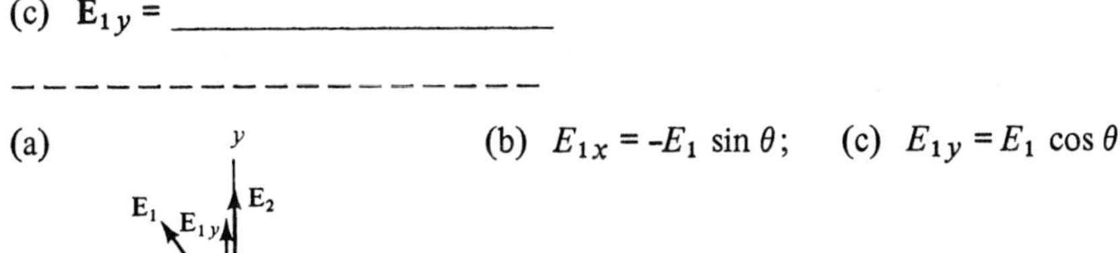

(a)

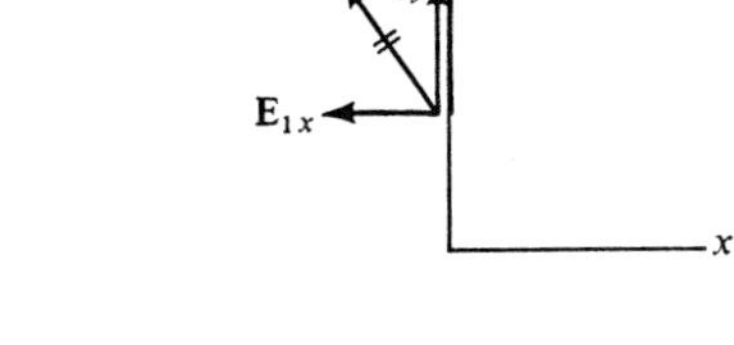

(b) $E_{1x} = -E_1 \sin\theta$; (c) $E_{1y} = E_1 \cos\theta$

1.6 Because θ is not explicitly given in the problem we need to express the $\sin\theta$ and $\cos\theta$ terms using the variables a and b which are given. In terms of the lengths a and b

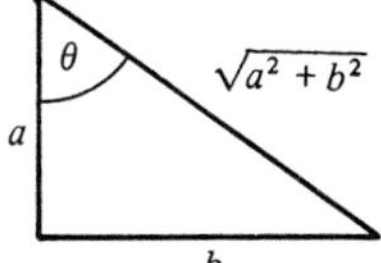

(a) $\cos\theta$ = ____________________

(b) $\sin\theta$ = ____________________

(a) $\cos\theta = \dfrac{a}{\sqrt{a^2+b^2}}$; (b) $\sin\theta = \dfrac{b}{\sqrt{a^2+b^2}}$

1.7 Substituting the answers of frame 1.6 into the answers of frame 1.5 we have

$$E_{1x} = (-E_1)\left(\frac{b}{\sqrt{a^2+b^2}}\right) \qquad E_{1y} = (E_1)\left(\frac{a}{\sqrt{a^2+b^2}}\right)$$

Having resolved E_1 we can now find the components of the resultant vector ($\mathbf{E}_1 + \mathbf{E}_2$). Algebraically

(a) E_{Total} in the y direction = ____________________

(b) E_{Total} in the x direction = ____________________

(a) $E_{Tx} = -\dfrac{1}{4\pi\epsilon_0}\left(\dfrac{q_1}{a^2+b^2}\;\dfrac{b}{\sqrt{a^2+b^2}}\right);$

(b) $E_{Ty} = \dfrac{1}{4\pi\epsilon_0}\left(\dfrac{q_1}{a^2+b^2}\;\dfrac{a}{\sqrt{a^2+b^2}} + \dfrac{q_2}{a^2}\right)$

1.8 The problem is essentially finished. You will remember from mechanics that if you know the components of a vector then you know everything about it. If you wish to finish the problem by doing the arithmetic, please do so. Compare your answer to the one given below. The data of the problem is in frame 1.3.

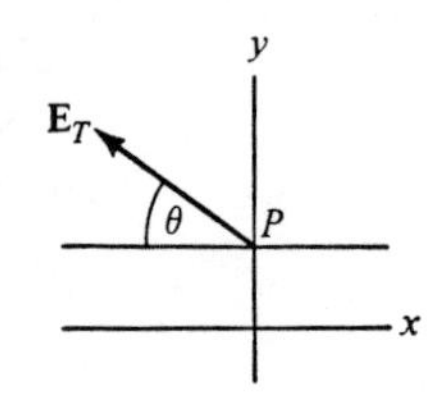

(a) E_{Tx} = 

(b) E_{Ty} = ____________

(c) θ = ____________

(a) $E_{Tx} = -128$ nt/coul; (b) $E_{Ty} = 172$ nt/coul;
(c) $\tan\theta = 172/128$, $\theta = 53^\circ$

Problem 2

A proton is accelerated from rest in a uniform electric field of strength 2.5×10^5 nt/coul. If the proton is accelerated over a distance of 0.4 m, what is the final kinetic energy and speed of the electron?

2.1 A crucial element in determining the motion of a particle is to "find the force." The force $\mathbf{F} = q\mathbf{E}$ acting on the charge q in an electrostatic field $\mathbf{E}$ is an example of "finding a force." We can thus discuss the motion of charged particles in electric fields.

Consider a proton (mass m and charge q) which enters a region of uniform electric field **E**. The proton is initially at rest. For convenience we think of the proton as entering through a hole.

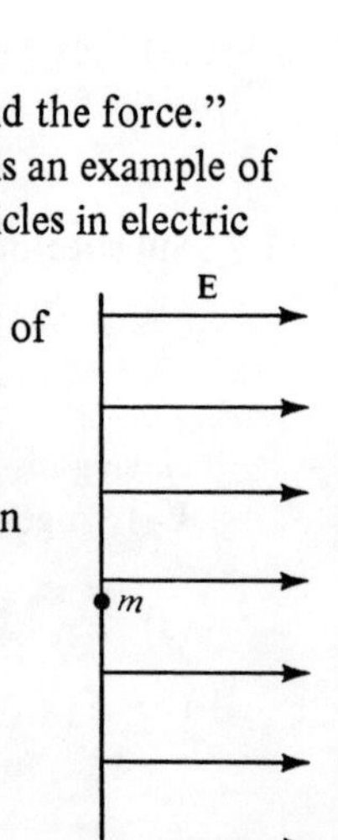

(a) Show by a vector the force **F** acting on the proton while it is in the field **E**.
(b) What will be the magnitude of this force?
(c) What does a "uniform electric field" mean?

- - - - - - - - - - - - - - - - - -

(a) 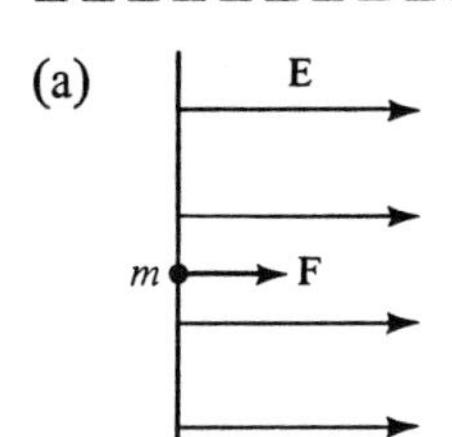

The electric force is in the direction of **E** for positive charges.

(b) $F = qE$ (where q is the proton charge and E is the electric field strength)

(c) The electric field **E** has the same magnitude and direction everywhere. Therefore, here **F** is constant.

2.2 Considering this a one-dimensional problem and ignoring gravitational effects, write Newton's second law of motion for the proton.

m ● → $\mathbf{F} = q\mathbf{E}$, axis x

- - - - - - - - - - - - - - - - - -

$qE = ma_x$

2.3 Since q, E, and m are constant in this problem the acceleration a_x is a constant. Algebraically, what will be the velocity of the proton after a time t, assuming that it starts from rest?

- - - - - - - - - - - - - - - - - -

$v_x = a_x t$ (The initial velocity was given to be zero.) This is a familiar kinematic expression for constant acceleration.

2.4 For uniformly accelerated motion starting from rest, the average speed during the time t is

$$v = \frac{1}{2} v_x$$

where v_x is given in the previous answer. From this fact the distance ℓ that the proton goes in a time t is

$\ell =$ __________

- - - - - - - - - - - - - - - - - -

$\ell = \frac{1}{2} v_x t$

2.5 Eliminate t by using the answers to frames 2.3 and 2.4.

$v_x = a_x t$
$\ell = \frac{1}{2} v_x \ell$
$v_x{}^2 = 2a_x \ell$

2.6 Hopefully you recognize this expression as a kinematic formula. Rewrite $v_x{}^2 = 2a_x \ell$ using a_x from the answer to frame 2.2.

$$v_x{}^2 = 2\frac{qE}{m}\ell$$

2.7 Rewriting the previous answer we have

$$\frac{1}{2} m v_x{}^2 = qE\ell$$

(a) What do we call the left side?
(b) What do we call the right side?

(a) Kinetic energy; (b) Work done by the electric field (force qE times distance ℓ)

2.8 We say that the proton has acquired kinetic energy. Devices which do this job are called accelerators. A TV picture tube operates in this manner with the particle being an electron.

A proton (mass = 1.67×10^{-27} kg, charge = 1.6×10^{-19} coul) is accelerated from rest through an electric field of strength 2.5×10^5 nt/coul for a distance 0.4 m. Compute the following. (Consider the case as shown in frame 2.1.)

(a) proton's final kinetic energy = ____________________

(b) proton's final speed = ____________________

(a) 1.6×10^{-14} joules
(b) 4.4×10^6 m/sec

Problem 3

An uncharged thin spherical metallic shell of radius $r = 5$ cm has a point charge $q = 3 \times 10^{-6}$ coul at the center of the shell. Use Gauss's law to determine the electric field inside and outside the metallic sphere. Determine numerically the field at a point 10 cm from the central charge.

3.1 Gauss's Law requires that the flux of the electric field through *any* closed surface be proportional to the net charge enclosed by *that* surface. The utilitiy of Gauss's law in problem solving derives from the fact that many situations have a symmetry which simplifies calculating the flux.

Ignore the metallic shell for a moment and qualitatively describe the electric field in the space around q.

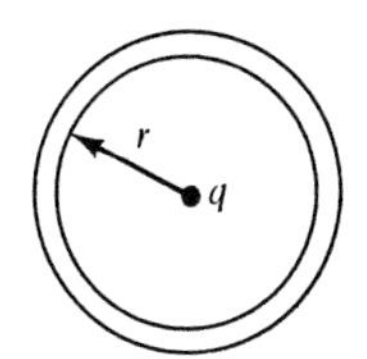

Directed radially outward

3.2 We are at liberty to choose any closed surface we wish.

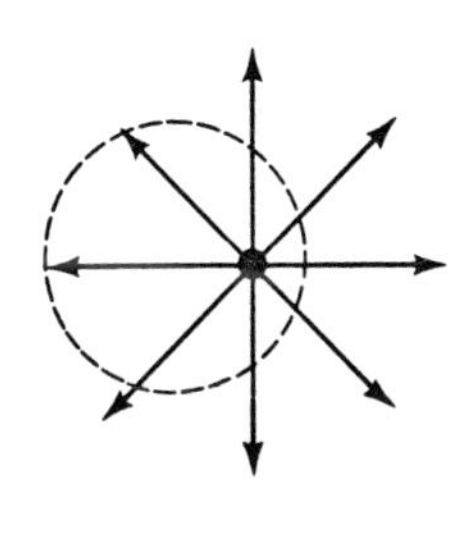

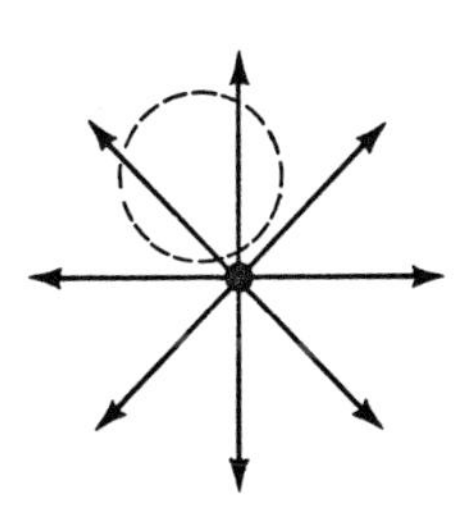

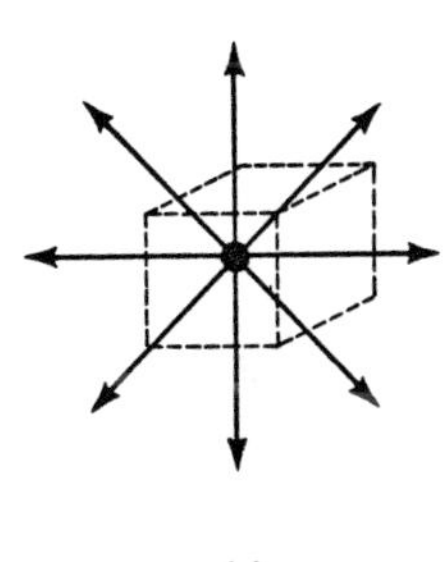

(a) (b) (c)

Above we represent the electric field of q by radially directed field lines. The dotted lines represent closed surfaces; two are spherical and one is cubical. According to Gauss's law which closed surfaces would have a net flux for the electric field?

(a) and (c) In (b) the surface encloses no net charge, therefore the electric flux is zero.

3.3 In (b) of frame 3.2, is the electric field inside the closed surface zero?

No. Gauss's law relates the flux ($\Sigma \mathbf{E} \cdot \Delta \mathbf{S}$) of the electric field to the charge. It doesn't relate the electric field alone.

3.4 In (b) of frame 3.2, is the electric field the same at all places on the surface of the imaginary sphere?

No. Field lines represent strong fields when they are close together. The points on the surface far away from q correspond to weaker fields than those closer to q.

3.5 We see from the two previous frames that some care in selecting a proper Gaussian surface is important.

- If you wish to know the field due to charges, the Gaussian surface must include those charges. Otherwise, all you know is that the flux of **E** is zero.
- If you choose a Gaussian surface carefully you can arrange to have the electric field a constant (including zero) for all parts of the surface.

Sketch in a Gaussian surface which satisfies the two criteria above.

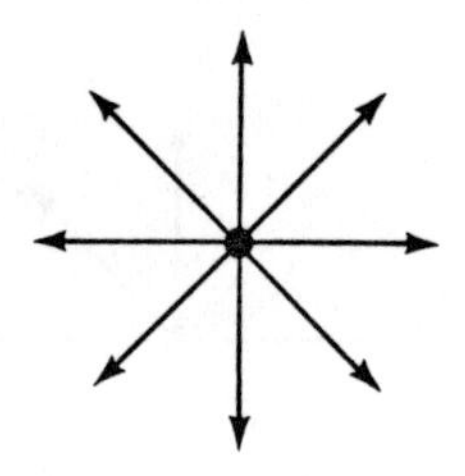

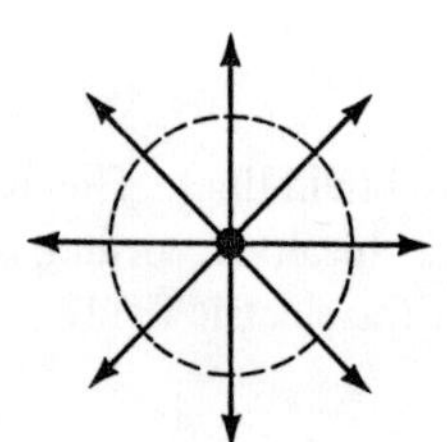

A spherical surface concentric with the charge satisfies the two criteria. The charge is enclosed by the surface and the electric field is constant at the surface.

3.6 What is the spatial relationship between the electric field (represented by radial lines) and elements of area $\Delta \mathbf{S}$ of the previous answer?

Parallel

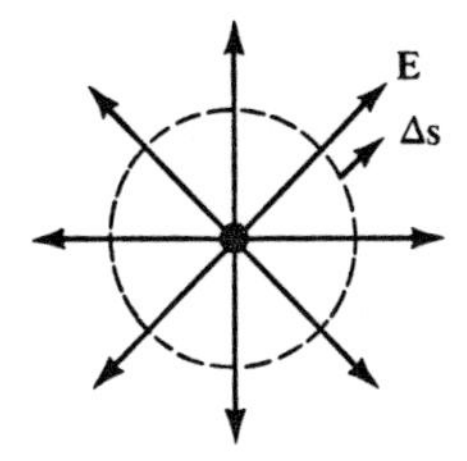

Elements of area **ΔS** are normal to the surface and are conventionally directed outward.

3.7 The equation $\Sigma \mathbf{E} \cdot \Delta \mathbf{S} = q/\epsilon_0$ means

$$E_1 \Delta S_1 \cos\theta_1 + E_2 \Delta S_2 \cos\theta_2 + \cdots = \frac{q}{\epsilon_0}$$

The summation is to be made over the entire closed surface.

If we let **E** and **ΔS** shown in the answer to frame 3.6 be the first term in the summation, what is $\cos\theta_1$?

---

1 (one) (Flux of **E** means the component of **E** in the direction **ΔS**. In the answer to frame 3.6 **E** and **ΔS** are parallel, so $\theta = 0°$; $\cos 0° = 1$.)

3.8 The angles θ_i in the summation will all be 0°, so we are simplifying the result.

$$E_1 \Delta S_1 + E_2 \Delta S + \cdots = \frac{q}{\epsilon_0}$$

In the sketch $\Delta \mathbf{S}_1$ has the same area as $\Delta \mathbf{S}_2$. Is $\mathbf{E}_1$ the same as $\mathbf{E}_2$?

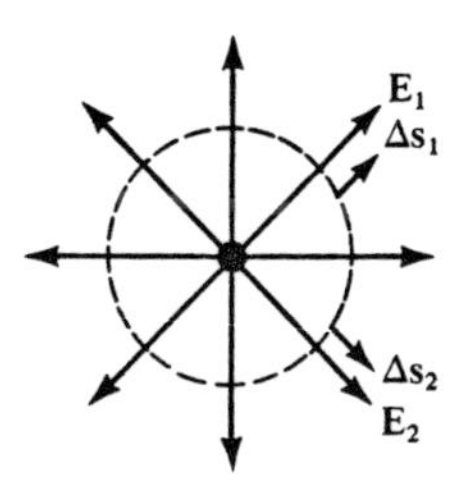

Yes. This, of course, is why we pick a Gaussian surface that is concentric. **E** is the same in magnitude everywhere on the Gaussian surface shown.

3.9 The summation is now even simpler.

$$E(\Delta S_1 + \Delta S_2 + \text{all the other } \Delta S\text{'s}) = q/\epsilon_0$$

In words, what does one call the term in parentheses?

The total area of the Gaussian surface

3.10 We arrive at the result of applying Gauss's law to a point charge.

$\Sigma \mathbf{E} \cdot \Sigma \mathbf{S} = \frac{q}{\epsilon_0}$ (Gauss's law)

$E \cdot \Sigma \Delta S = \frac{q}{\epsilon_0}$ (Because $\mathbf{E}$ and $\Delta \mathbf{S}$ are everywhere parallel and $\mathbf{E}$ is constant at the surface)

$E \cdot 4\pi r^2 = \frac{q}{\epsilon_0}$ (Because the area of the sphere is $4\pi r^2$)

What is the expression for E a distance r away from a point charge?

$$E = \frac{q}{4\pi\epsilon_0 r^2}$$

Note that the application of Gauss's law permitted the determination of the electric field at the points corresponding to the radius of the Gaussian surface.

3.11 Would the result be any different from

$$E = \frac{q}{4\pi\epsilon_0 r^2}$$

if we applied Gauss's law where an outside spherical shell is shown with an excess charge?

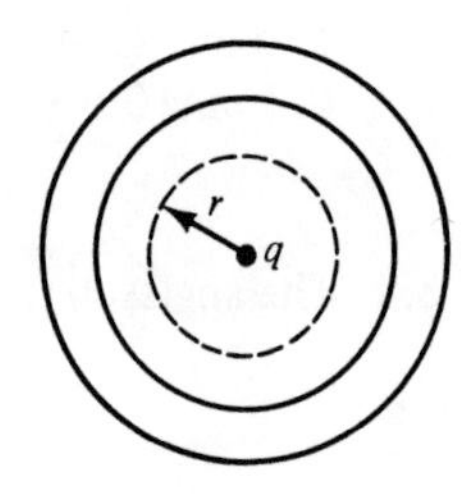

No. Gauss's law only includes the net charge inside the Gaussian surface.

3.12 Part of the problem is completed. The field inside the spherical shell is

$$E = \frac{q}{4\pi\epsilon_0 r^2}$$

where q is the point charge and r is a value up to the radius of the spherical shell.

Consider now applying Gauss's law using a spherical surface as shown to the right. The Gaussian surface is inside the uncharged conductor. The electric field E in the conductor in the static situation is zero. Otherwise, mobile charges would rearrange themselves under the influence of the electric field. What must be the flux of $\mathbf{E}$ through the Gaussian surface in the metal?

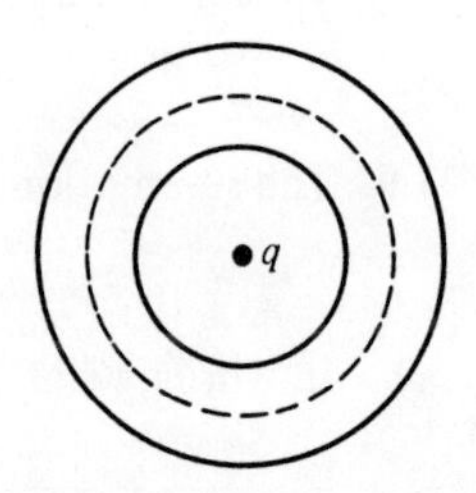

Zero (If $\mathbf{E}$ is zero then $\Sigma \mathbf{E} \cdot \Delta \mathbf{S} = 0$.)

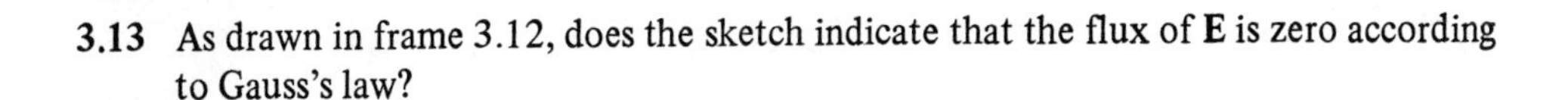

3.13 As drawn in frame 3.12, does the sketch indicate that the flux of **E** is zero according to Gauss's law?

No (As shown, the net charge enclosed is q; therefore, $\Sigma \mathbf{E} \cdot \Delta \mathbf{S} = q/\epsilon_0$, the flux of **E** is not zero.)

3.14 We must arrange things to meet the physical fact that the electric field (and thus also the flux through the surface) is zero inside the metal. What do we have to do to the net charge to satisfy this physical requirement according to Gauss's law?

Require that the net charge enclosed be zero.

3.15 Schematically we can fix things up if we arrange to have negative charges on the inner shell. Then we have

$$\Sigma q = \underbrace{+q}_{\text{at the center}} \quad \underbrace{-q}_{\text{on inner shell}} = 0$$

This means the net charge inside the Gaussian surface as shown in the sketch is zero.

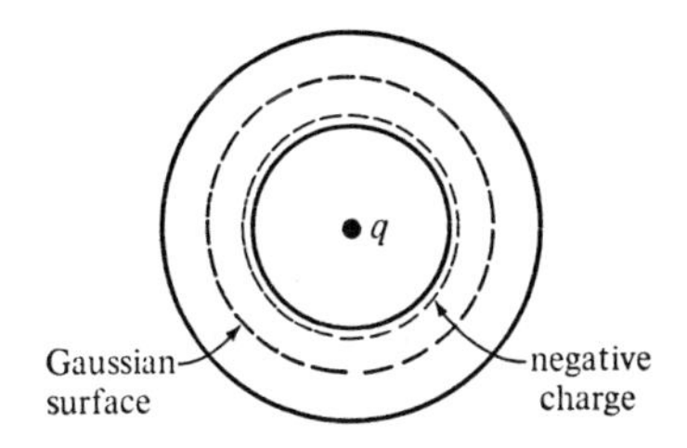

Everything is now all right analytically except where do the negative charges come from? In the problem it is stated that the spherical conductor has no excess charge. The metal conductor does have charges (electrons) which can move around a bit. We say that the negative charges on the inside of the shell are *induced* due to the presence of the central charge q.

To maintain the net charge of zero on the shell we must indicate an equal and opposite induced charge on the outside surface. The net charge is thus zero.

If you draw a concentric Gaussian surface whose radius is greater than that of the shell, what will be the net charge enclosed?

$+q$

The equal and opposite induced charges do not contribute to the net charge enclosed. In this problem the electric field inside and outside the sphere is such that the spherical uncharged shell has no effect.

3.16 Determine numerically the electric field at $R = 10$ cm.

27×10^5 nt/coul

$$E = \frac{1}{4\pi\epsilon_0}\frac{q}{r^2}$$

$$E = \frac{9 \times 10^9 \times 3 \times 10^{-6}}{1 \times 10^{-2}}$$

$$E = 27 \times 10^5 \text{ nt/coul}$$

Problem 4

Consider the cubical Gaussian surface shown in the diagram. The electric field in the coordinate system is

$$E_x = bx^{\frac{1}{2}}, E_y = 0, E_z = 0$$

Given that the cube measures 0.1 m on a side and $b =$ 800 nt/coul-m$^{\frac{1}{2}}$, how much charge is inside the cube?

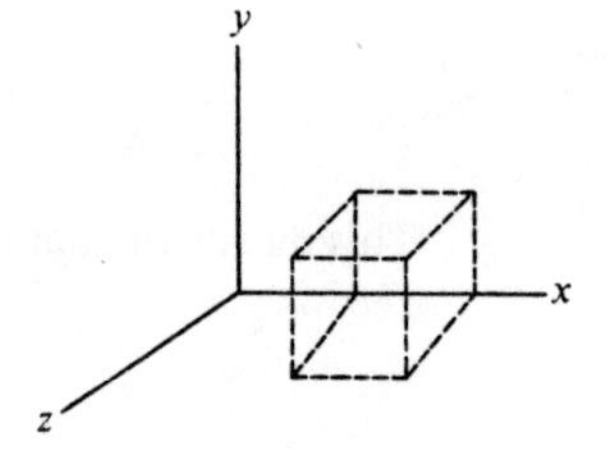

4.1 The left face of the cube shown is a distance a away from the origin. Each side of the cube is also a in length. An electric field given by

$$E_x = bx^{\frac{1}{2}}, E_y = 0, E_z = 0$$

is present. We wish to calculate the flux of **E** through the cube and use this result to obtain the charge enclosed by the cube. Qualitatively describe the given electric field.

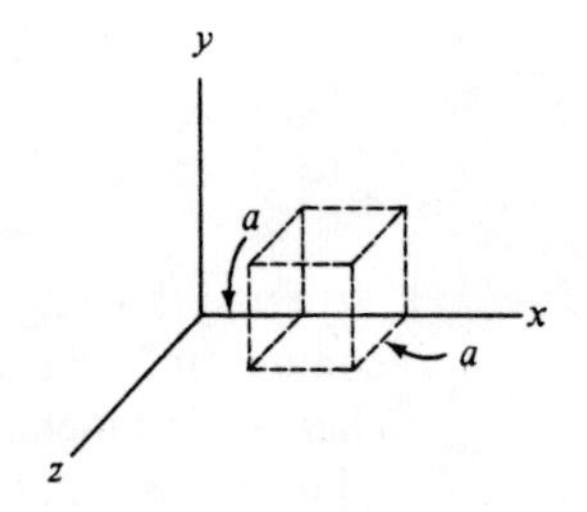

The field is in the x direction only. The strength of the field becomes larger as you go away from the origin in the direction of x.

4.2 Will electric field lines pass through all faces of the cubical closed surface? Explain.

No. Since the electric field is only in the x direction, field lines will be parallel to all faces except the left and right faces.

4.3 Will the electric field have the same intensity at all surface elements ΔS of the left face of the cube?

Yes. All surface elements of the left side are at the same value of x and the electric field has a fixed intensity at any particular x (i.e., $E_x = bx^{\frac{1}{2}}$).

4.4 Is the electric field intensity the same for every surface element of the right face of the cube?

Yes (same reason as the previous answer)

4.5 Is the electric field intensity the same at the left face as it is at the right face?

No

$E = bx^{\frac{1}{2}}$

At the left face $x = a$, while at the right face $x = 2a$.

4.6 For $b = 800$ nt/coul-m$^{\frac{1}{2}}$ and $a = 0.1$ m, calculate the following.

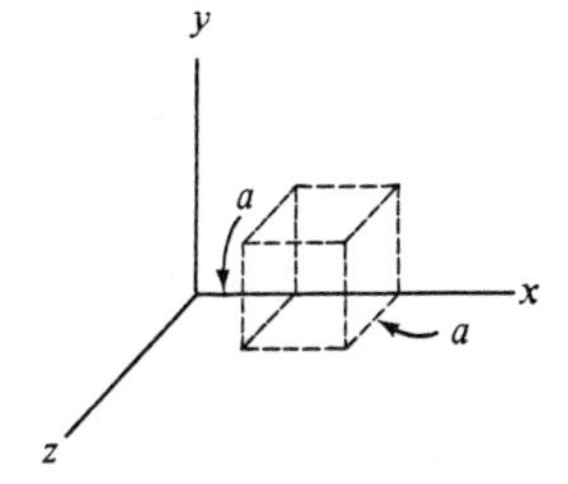

E_L = __________ nt/coul at left face

E_R = __________ nt/coul at right face

$E_L = bx^{\frac{1}{2}}$

$E_L = (800 \text{ nt/coul-m}^{\frac{1}{2}}) \times (0.1 \text{ m})^{\frac{1}{2}}$

$E_L = 250$ nt/coul

Similarly for $x = 2a = 0.2$ m, $E_R = 360$ nt/coul

4.7 We have already established that the total flux through the closed cube involves only the left and right sides.

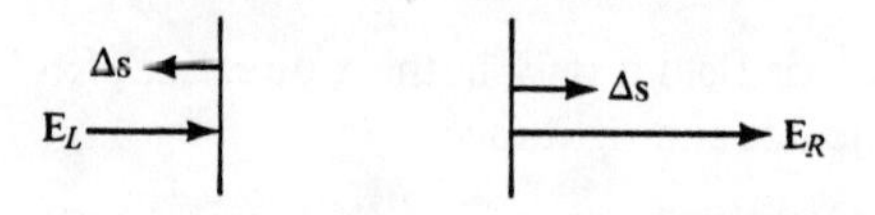

$$\Phi_E = \Sigma \mathbf{E} \cdot \Delta \mathbf{S} = \Sigma_L \mathbf{E}_L \cdot \Delta \mathbf{S} + \Sigma_R \mathbf{E}_R \cdot \Delta \mathbf{S}$$

Rewrite this equation in the correct scalar form (i.e., consider the orientation of the relevant vectors).

$\Phi_E = -\Sigma_L E_L \Delta S + \Sigma_R E_R \Delta S$

The minus sign because cos 180° = -1 and the plus sign because cos 0° = +1.

4.8 In view of frames 4.6 and 4.7 concerning the value of E_L and E_R we can write

$$\Phi_E = -E_L \Sigma_L \Delta S + E_R \Sigma_R \Delta S$$

where (a) $\Sigma_L \Delta S =$ ____________________

(b) $\Sigma_R \Delta S =$ ____________________

(a) $a^2 = 0.01 \text{ m}^2$; (b) $a^2 = 0.01 \text{ m}^2$

4.9 From frame 4.6 we have $a = 0.1$ m, $E_L = 250$ nt/coul, and $E_R = 360$ nt/coul, so

$\Phi_E =$ ____________________

Obtain a number.

1.1 nt-m²/coul

$$\Phi_E = -E_L \Sigma_L \Delta S + E_R \Sigma_R \Delta S$$
$$\Phi_E = -E_L a^2 + E_R a^2$$
$$\Phi_E = a^2(-E_L + E_R)$$
$$\Phi_E = 0.01 \text{ m}^2(-250 \text{ nt/coul} + 360 \text{ nt/coul})$$
$$\Phi_E = 1.1 \text{ nt-m}^2/\text{coul}$$

4.10 From Gauss's law we can now determine the charge enclosed.

$$\Sigma_{\text{cube}} \mathbf{E} \cdot \Delta \mathbf{S} = \frac{q}{\epsilon_0}$$

Use $\epsilon_0 = 8.85 \times 10^{-12}$ coul^2/nt-m^2 to find q.

$q = 9.7 \times 10^{-12}$ coul

Note that the net charge enclosed is positive since the flux of **E** was positive.

$q = \epsilon_0 \Phi_E$

$q = (1.1 \text{ nt-m}^2/\text{coul})(8.85 \times 10^{-12} \text{ coul}^2/\text{nt-m}^2)$

$q = 9.7 \times 10^{-12}$ coul

SELF–TEST

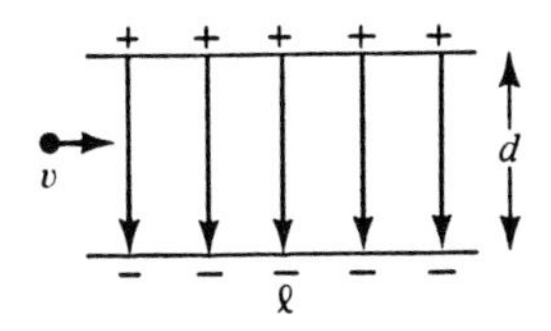

1 An electron enters halfway between two parallel plates of length ℓ. The entering velocity of the electron is 1.5×10^7 m/sec in the horizontal direction only. Given that $\ell = 2$ cm and $d = 1$ cm, determine the magnitude of the maximum electric field that can be produced between the plates such that the electron will just strike the end of the plate. The negative charge on the electron is 1.6×10^{-19} coul and the mass is 9×10^{-31} kg. Also calculate the potential difference across the plates to maintain the electric field.

2 Two like charges are separated by a distance $d = 20$ cm. The charge on one is twice the charge on the other. Determine the point between the charges at which the electric field is zero.

3 Two uniformly charged concentric spheres have radii of 50 cm and 75 cm. The inner and outer spheres have charges of $+6 \times 10^{-6}$ coul and -12×10^{-6} coul respectively. Calculate the magnitude and direction of the electric field at the following radii: 20 cm, 60 cm, and 90 cm.

Answers to Self-Test

1 $E = 3.3 \times 10^4$ nt/coul
$V = 330$ volts

2 11.7 cm from the larger charge

3 $r = 20$ cm, $E = 0$
$r = 60$ cm, $E = 15$ nt/coul, directed radially outward
$r = 90$ cm, $E = 15$ nt/coul, directed radially inward

CHAPTER THREE
Electric Potential

If the sample problems and objectives below identify your weak points, go directly to the programmed study section on page 41. If not, try the problems and compare your answers with those that follow. If you can do all the problems easily and if you are familiar with the objectives, you may wish to skip all or part of this chapter. The programmed study section covers techniques and concepts basic to solving the sample problems and fulfilling the objectives in this chapter. A programmed, step-by-step solution of each sample problem begins on page 47. A self-test is included at the end of the chapter.

SAMPLE PROBLEMS AND OBJECTIVES

Problem 1

Determine the electric potential at a point halfway between the two charges.

$d = 2\text{m}$

$q_1 = 3 \times 10^{-6}$ coul $\quad q_2 = 2 \times 10^{-6}$ coul

Objectives:

1. Defining a physically sensible zero of electric potential.
2. Reviewing the electric fields of point charges.
3. Discussing the additivity of electric potentials.

Problem 2

A pair of parallel plates are charged as shown, thus establishing a constant electric field E in the space between the plates. The electric field has a magnitude of 1.25×10^4 nt/coul and the separation of the plates is 2.0 cm. Each plate has an area of 24 cm^2. The magnitude of the charge on each plate is 2.5×10^{-8} coul. Calculate the potential difference between the plates.

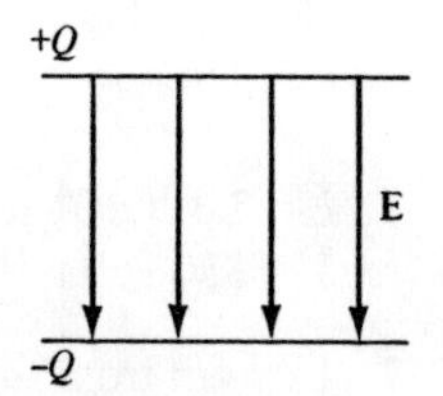

Objectives:

1. Reviewing the meaning of potential difference in a physical context.
2. Discussing the properties of capacitors.

Problem 3

An insulated conducting sphere of radius 20 cm is given a charge of 5×10^{-8} coul. What is the electric potential at the surface of the sphere?

Objectives:
1. Reviewing Gauss's law.
2. Discussing the size of electric potentials in electrostatic situations.

Answers to Sample Problems

See page 47 for programmed, step-by-step solutions to these problems.

Problem 1

$V = 9 \times 10^3$ volts

Problem 2

$V_B - V_A = 2.5 \times 10^2$ volts

Problem 3

$V_B = 2.26 \times 10^3$ volts

PROGRAMMED STUDY SECTION

1 In actual practice electric fields are seldom measured directly. In the television picture tube, for example, the brightness (florescence) on the screen is the result of energetic electrons colliding with phosphor sprayed onto the glass screen. These energetic electrons are produced by acceleration in electric fields. However, advertisements always refer to the electron beams as "25,000 volt electron beams" rather than "2.5 $\times 10^5$ nt/coul electrostatically accelerated beams."

Describe qualitatively what would happen to a positive test charge q_0 if it were placed, but not held, in the electric field of a positive charge q which is fixed in space.

q_0

$+q$

The test charge would be accelerated in the direction of the electric field of $+q$. In this case the test charge would accelerate away from q. (You probably think of this as "like charges repel each other.")

2 As the previous answer indicates the acceleration of q_0 is "away" from $+q$. What is the direction of the resultant force on the test charge q_0?

As always, the acceleration and resultant force are in the same direction.

3 Draw the following forces on q_0 to represent those forces necessary to keep q_0 in equilibrium.

(a) the electrical force
(b) an external force (say of a man)

q_0

$+q$

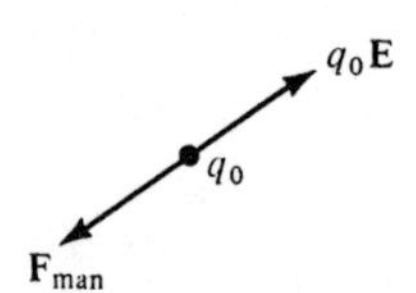

4 In the previous answer the electrical force is labeled q_0**E**. What is the source of **E**?

The charge $+q$

5 How are the force vectors $\mathbf{F}_{man}$ and $q_0\mathbf{E}$ related in the answer to frame 3?

Equal magnitude and oppositely directed ($\mathbf{F}_{man} = -q_0\mathbf{E}$ or $\mathbf{F}_{man} + q_0\mathbf{E} = 0$ since q_0 is in equilibrium.)

6 The force $\mathbf{F}_{man}$ has been introduced so that the test charge can be used to explore the electric field of $+q$. We have already seen that q_0 would accelerate away if it were "free."

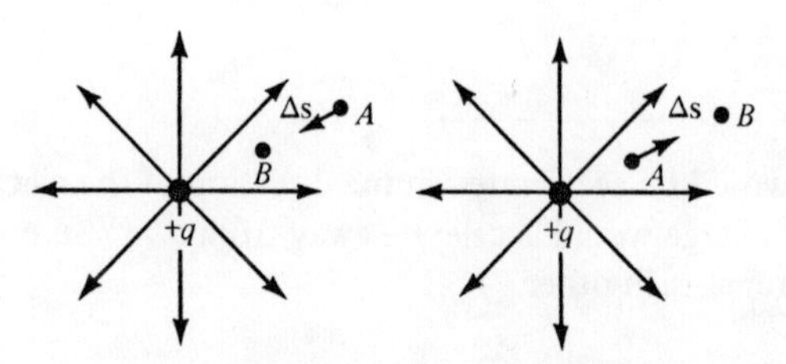

On page 42 are shown two identical charges. In both cases we wish to calculate the work done by the external force $\mathbf{F}_{man}$ in going from A to B under the condition that at all times $\mathbf{F}_{man} = -q_0\mathbf{E}$. Let us take the simple radial path.

(a) In both cases $\mathbf{F}_{man}$ is the same along all points of the path from A to B. (True/False)
(b) The displacement vector $\Delta \mathbf{s}$ goes from A to B in both cases. (True/False)
(c) In both cases $\mathbf{F}_{man}$ points in a direction from A to B. (True/False)

(a) False; since $F_{man} = q_0 E$, F is not constant because E is not constant.

$$E = \frac{1}{4\pi\epsilon_0} \frac{q}{r^2}$$

(b) True
(c) False; in both cases $\mathbf{F}_{man}$ points "toward" the $+q$ charge.

7 In view of answers (b) and (c) of the previous frame, what chief difference would there be in calculating the work done by the man for the two cases shown.

$$W = \sum_{A}^{B} \mathbf{F}_{man} \cdot \Delta \mathbf{s}$$

One would be positive and one would be negative.

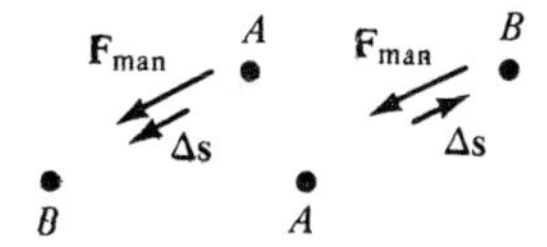

Positive, $\cos\theta = 1$

Negative, $\cos\theta = -1$

8 The answer of the previous frame involves the product of force and distance. We call such products ________________

----

work (This is the work done by the man in moving q_0 from A to B.)

9 Complete the following from our example of frame 6.

(a) (Positive/Negative) work is done by the man in moving a test charge against electric field lines.

(b) (Positive/Negative) work is done by the man in moving a test charge in the direction of electric field lines.

(a) Positive;
(b) Negative

10 The result noted in the previous frame is summarized as follows for a positive charged particle in an electric field.

- If an external agent (say the man) does positive work in going from A to B then we say that B is at a *higher* electric potential (or similarly A is at a *lower* electric potential than B).
- If an external force does negative work in going from A to B then B is at a *lower* electric potential (or similarly A is at a *higher* electric potential than B).

11 Imagine that a man supplies an external force such that $\mathbf{F}_{\text{man}} = -q_0\mathbf{E}$ in the three examples below.

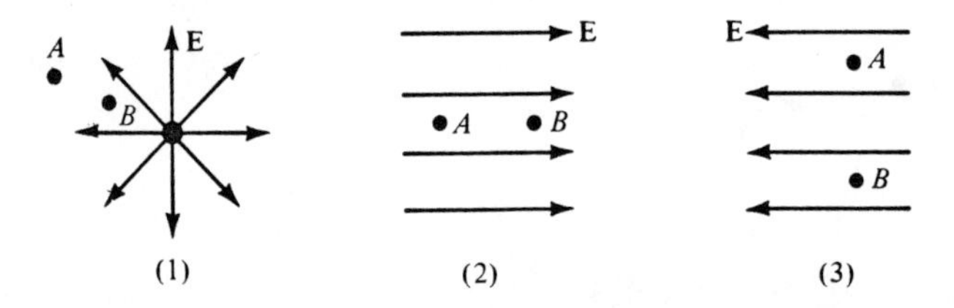

Complete the following statements in terms of whether

$$\sum_{A}^{B} \mathbf{F}_{\text{man}} \cdot \Delta \mathbf{s}$$

is either positive or negative.

(a) In (1) A is at a (lower/higher) electric potential than B.
(b) In (2) A is at a (lower/higher) electric potential than B.
(c) In (3) A is at a (lower/higher) electric potential than B.

(a) lower; (b) higher; (c) both at the same potential

Note:

$\mathbf{F}_{man}$, θ, Δs

$\cos\theta = 0; \mathbf{F}_{man} \cdot \Delta s = 0$

When the work is zero in going from A to B, then A and B are at the same electric potential.

12 We can now write down formally that

$$W_{AB} = \sum_{A}^{B} \mathbf{F}_{man} \cdot \Delta s$$

as the work done by an outside agent in moving a test charge q_0 from point A to point B. Since $\mathbf{F}_{man} = -q_0\mathbf{E}$ we can write

$$W_{AB} = -q_0 \sum_{A}^{B} \mathbf{E} \cdot \Delta s$$

Dividing by q_0 we have

$$\frac{W_{AB}}{q_0} = -\sum_{A}^{B} \mathbf{E} \cdot \Delta s$$

The quantity on the right side of the above equation will be

(a) (positive/negative) when Δs is in the direction of **E**

(b) (positive/negative) when Δs is opposite the direction of **E**

(c) ______________ where **E** and Δs are perpendicular

(a) negative $(-E\Delta s)$; (b) positive $[-(-E\Delta s)]$; (c) zero

13 The left side of the equation

$$\frac{W_{AB}}{q_0} = -\sum_{A}^{B} \mathbf{E} \cdot \Delta \mathbf{s}$$

can thus be positive, negative, or zero.

(a) If positive work is done in going from A to B, then B is at a (higher/lower) electric potential than A.
(b) If negative work is done in going from A to B, then B is at a (higher/lower) electric potential than A.

(a) higher; (b) lower

14 The work per unit charge in going from A to B is called the electric potential difference ($V_B - V_A$) so the equation becomes

$$V_B - V_A = -\sum_{A}^{B} \mathbf{E} \cdot \Delta \mathbf{s}$$

If the value ($V_B - V_A$) is independent of the actual path between A and B, then **E** is called a ______________________ field.

----

conservative (All electrostatic fields have this property. See your textbook for a further discussion of conservative fields. The earth's gravitational field is also conservative.)

SOLUTIONS TO SAMPLE PROBLEMS

Problem 1

Determine the electric potential at a point halfway between the two charges.

|←— d = 2m —→|

$q_1 = 3 \times 10^{-6}$ coul $q_2 = 2 \times 10^{-6}$ coul

1.1 Because of the mathematical complications involved in problems of this type we will have to resort to a kind of tricky approximation. However, the result is important and along the way you will hopefully get some better feeling for the concept of electric potential.

As discussed in the programmed study section

$$V_B - V_A = -\sum_{A}^{B} \mathbf{E} \cdot \Delta \mathbf{s}$$

defines the potential difference between two points in an electric field.

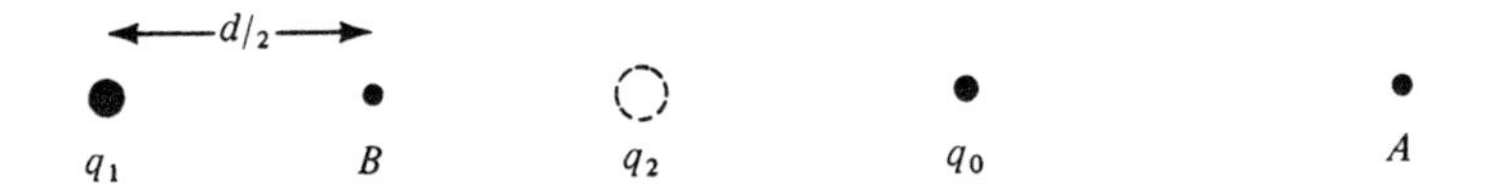

We will ignore q_2 and calculate the potential at the point of interest due to q_1 alone. To make things easier we will arrange the path to be straight.

The idea of electric potential derives from the fact that one must do work to move a test charge q_0 from place to place in an electric field. Show by arrows on the diagram above **E** due to q_1 at the site of q_0 and Δs of q_0 in moving from A to B.

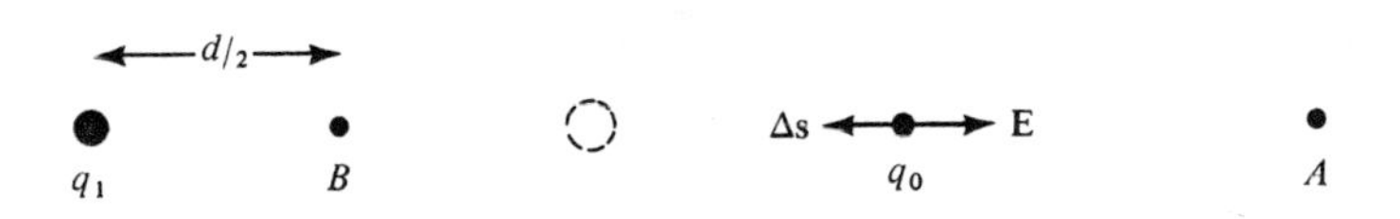

1.2 In the absence of any other force (other than $q_0\mathbf{E}$), will q_0 move in the direction Δs as shown? (Remember q_0 is a positive charge.)

No (The test charge would naturally move in the direction of **E**.)

1.3 To move q_0 as shown an external force must be supplied in the direction of $\Delta\mathbf{s}$. Clearly, then, this external force must do positive work. $\Delta W = \mathbf{F} \cdot \Delta\mathbf{s}$ is positive if $\mathbf{F}$ and $\Delta\mathbf{s}$ are in the same direction. If we must do this positive work we are moving toward a position of (higher/lower) potential.

---

higher $(V_B > V_A)$

1.4 We know that

$$V_B - V_A = -\sum_A^B \mathbf{E} \cdot \Delta\mathbf{s}$$

must be positive. What is the angle between $\mathbf{E}$ and $\Delta\mathbf{s}$ in the answer to frame 1.1?

180°

1.5 Rewrite the formula

$$V_B - V_A = -\sum_A^B \mathbf{E} \cdot \Delta\mathbf{s}$$

using the information of the last frame.

$$V_B - V_A = \sum_A^B E\Delta s$$

$$\begin{aligned} -\sum_A^B \mathbf{E} \cdot \Delta\mathbf{s} &= -\sum_A^B E\Delta s \cos\theta \\ &= -\sum_A^B E\Delta s \cos 180^\circ \\ &= -\sum_A^B E\Delta s(-1) \\ &= \sum_A^B E\Delta s \end{aligned}$$

1.6 Unlike the situation described in the second problem of this chapter, **E** is not constant. Write an expression for the electric field of a point charge like q_1.

$E = \frac{1}{4\pi\epsilon_0} \frac{q}{r^2}$, directed radially from the charge

1.7 Determing the sum of terms in

$$\sum_A^B E \Delta s$$

requires that we resort to an approximation. The reason for the difficulty is that **E** is different for each Δs. We have already considered directions so we need only consider magnitudes.

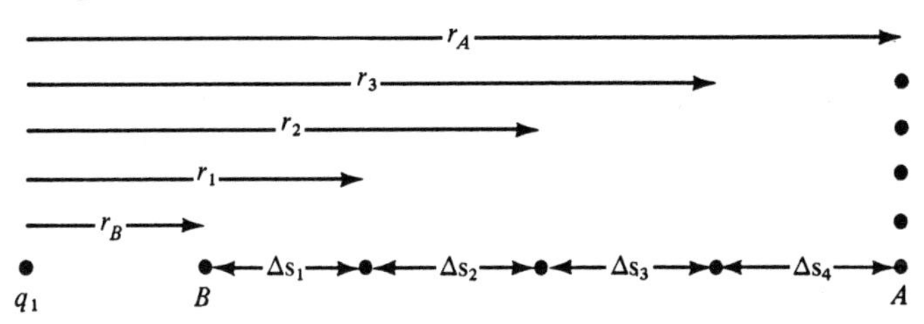

Above we break up the path from A to B into displacements labeled Δs_1, etc. If $\Delta s_1 = r_1 - r_B$, compute the following.

(a) $\Delta s_2 =$ ____________

(b) $\Delta s_3 =$ ____________

(c) $\Delta s_4 =$ ____________

(a) $\Delta s_2 = r_2 - r_1$; (b) $\Delta s_3 = r_3 - r_2$; (c) $\Delta s_4 = r_A - r_3$

1.8 Compute the following, again referring to the diagram in frame 1.7.

$$E \text{ at } r_A = \frac{1}{4\pi\epsilon_0} \frac{q_1}{{r_A}^2}$$

(a) E at $r_1 =$ ____________

(b) E at $r_2 =$ ____________

(c) E at $r_3 =$ ____________

(d) E at $r_B =$ ____________

(a) $\frac{1}{4\pi\epsilon_0}\frac{q_1}{r_1^2}$; (b) $\frac{1}{4\pi\epsilon_0}\frac{q_1}{r_2^2}$; (c) $\frac{1}{4\pi\epsilon_0}\frac{q_1}{r_3^2}$; (d) $\frac{1}{4\pi\epsilon_0}\frac{q_1}{r_4^2}$

1.9 Because E is not constant over Δs we must use an approximation to sum the terms.

$$E = \frac{1}{4\pi\epsilon_0}\frac{q_1}{r_3^2} \qquad E = \frac{1}{4\pi\epsilon_0}\frac{q_1}{r_A^2}$$

$$\Delta s_4 = r_A - r_3$$

Here we show one part of the diagram from frame 1.7. We imagine making Δs_4 (and all others) small enough so that $r_3 \approx r_A$. For the approximation $r_A^2 = r_A r_3$, what is r_3^2?

$r_A r_3$

1.10 Now we can write out all terms in the sum

$$\sum_A^B E\Delta s$$

We imagine that there are only the four represented in the diagram of frame 1.7. Let us write the first term in going from r_A to r_3 where we approximate

$$E = \frac{1}{4\pi\epsilon_0}\frac{q_1}{r_A r_3}$$

For that term

$$\underbrace{\frac{1}{4\pi\epsilon_0}\frac{q_1}{r_A r_3}}_{E}\underbrace{(r_A - r_3)}_{\Delta s}$$

Now we can write this as

$$\frac{q_1}{4\pi\epsilon_0}\left(\frac{r_A - r_3}{r_A r_3}\right)$$

because the coefficient $q_1/4\pi\epsilon_0$ will be the same for all similar terms in the sum. Write the term in parentheses as the difference of two fractions.

$\left(\frac{1}{r_3} - \frac{1}{r_A}\right)$

$$\frac{r_A}{r_A r_3} - \frac{r_3}{r_A r_3} = \frac{1}{r_3} - \frac{1}{r_A}$$

We just subtract fractions having the same denominator and divide out the common terms in each fraction.

1.11 After all this we have

$$V_B - V_A = \frac{q_1}{4\pi\epsilon_0}\left(\frac{1}{r_3} - \frac{1}{r_A}\right) + \text{other similar terms}$$

For the four Δs's previously shown we have

$$V_B - V_A = \frac{q_1}{4\pi\epsilon_0}\left[\left(\frac{1}{r_3} - \frac{1}{r_A}\right) + \left(\frac{1}{r_2} - \frac{1}{r_3}\right) + \left(\frac{1}{r_1} - \frac{1}{r_2}\right) + \left(\frac{1}{r_B} - \frac{1}{r_1}\right)\right]$$

Combine terms.

$$V_B - V_A = \frac{q_1}{4\pi\epsilon_0}\left(\qquad\qquad\qquad \right)$$

$$V_B - V_A = \frac{q_1}{4\pi\epsilon_0}\left(-\frac{1}{r_A} + \frac{1}{r_B}\right)$$

All remaining terms cancel (for example, $\frac{1}{r_3} - \frac{1}{r_3} = 0$).

1.12 To interpret the answer we use physical reasoning. We let $r_A = \infty$ and $V_A = 0$. With these statements applied to the previous answer, compute V_B.

$$V_B = \frac{q_1}{4\pi\epsilon_0}\frac{1}{r_B}; \text{ as } V_A = 0, \frac{1}{r_A} = \frac{1}{\infty} = 0.$$

This is the general form for the electric potential at a distance r_B from the point charge. We use $V_A = 0$ at $r_A = \infty$ since the electric field is essentially zero when we are that far away from the charge.

1.13 In this particular problem $r_B = d/2$, so $V_B = q_1/2\pi\epsilon_0 d$ due to q_1. If we repeat the whole process for q_2, we obtain $V_B = -q_2/2\pi\epsilon_0 d$ due to the negative charge. The total electric potential at the midpoint is the sum of potentials due to each charge: $V_T = \Sigma V$'s.

$V_T =$ ______________________ at the mid-point of the two charges.

$V_T = \frac{1}{2\pi\epsilon_0 d}(q_1 - q_2)$, where now we use the only magnitudes of the charges when substituting for q_1 and q_2

1.14 Obtain a numerical value for V_T using data from the problem.

$V_T = 9 \times 10^3$ volts

$$V_T = \frac{1}{2\pi\epsilon_0 d} [q_1 - q_2]$$

$$V_T = \frac{1}{2\pi\epsilon_0 d} [(3 \times 10^{-6}) - (2 \times 10^{-6})]$$

$$V_T = \frac{10^{-6} \times 18 \times 10^9}{2}$$

$$V_T = 9 \times 10^3 \text{ volts}$$

Problem 2

A pair of parallel plates are charged as shown, thus establishing a constant electric field E in the space between the plates. The electric field has a magnitude of 1.25×10^4 nt/coul and the separation of the plates is 2.0 cm. Each plate has an area of 24 cm². The magnitude of the charge on each plate is 2.5×10^{-8} coul. Calculate the potential difference between the plates.

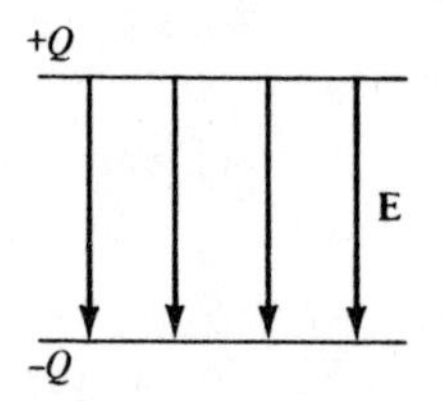

2.1 In this problem we have a constant electric field of magnitude 1.25×10^4 nt/coul. We want first to use the equation

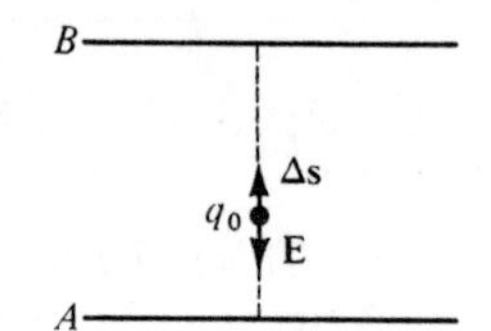

$$V_B - V_A = -\sum_A^B \mathbf{E} \cdot \Delta\mathbf{s}$$

to calculate the potential difference between the plates. In order for an external agent (say your hand) to move the positive test charge q_0 from A to B must you do positive or negative work? Remember

$$W_{\text{you}} = \Sigma \mathbf{F}_{\text{you}} \cdot \Delta\mathbf{s}_{\text{of } q_0}$$

Positive

The electric field would make the charge go from B to A so you must supply an opposite force **F**. Your force and the displacement of q_0 are parallel. Thus, the angle between **F** and Δs is zero.

$$\Delta W = \mathbf{F} \cdot \Delta\mathbf{s} = F\Delta s \cos 0° = +F\Delta s$$

2.2 As you must do positive work against the electric field, B is at a (higher/lower) electric potential than A.

---

higher

2.3 Thus, since $V_B > V_A$ you should obtain a positive number for $(V_B - V_A)$.

$$V_B - V_A = -\sum_{A}^{B} \mathbf{E}\cdot\Delta\mathbf{s}$$

In frame 2.1 what is the angle between **E** and **Δs** as q_0 goes along a perpendicular between the plates?

180°

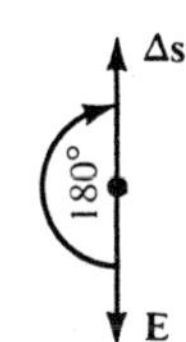

2.4 Rewrite the equation of the previous frame using the information in the answer.

$$V_B - V_A = +\sum_{A}^{B} \mathbf{E}\cdot\Delta\mathbf{s}$$

Note:

$$V_B - V_A = -\sum_{A}^{B} \mathbf{E}\cdot\Delta\mathbf{s}\cos 180°$$

$$V_B - V_A = -\sum_{A}^{B} -E\Delta s$$

$$V_B - V_A = +\sum_{A}^{B} E\Delta s$$

2.5 Does **E** depend on the position of q_0 between the plates?

No (**E** is constant.)

2.6 Since **E** is constant

$$V_B - V_A = E \sum_A^B \Delta s$$

What is $\sum_A^B \Delta s$?

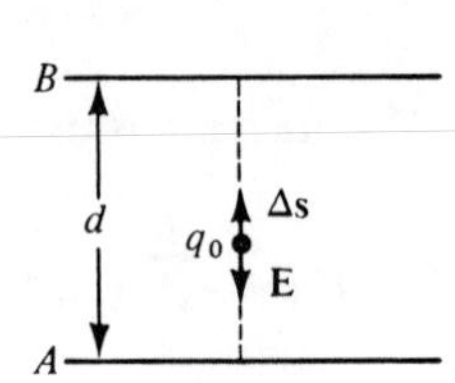

---

d

2.7 The answer then is

$$V_B - V_A = Ed$$

Obtain a numerical value for the potential difference between the parallel plates.

$E = 1.25 \times 10^4$ nt/coul
$d = 2$ cm

Don't mix units!

$V_B - V_A = 2.5 \times 10^2$ volts

$Ed = (1.25 \times 10^4 \text{ nt/coul})(2 \times 10^{-2} \text{ m})$
$Ed = 2.5 \times 10^2$ volts

2.8 This number is not an unusual one. Such an arrangement of plates is typical of accelerating anodes in oscilloscopes.

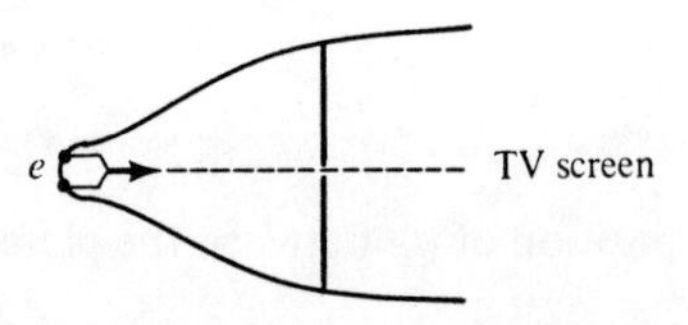

Electrons enter the region between the plates and are accelerated by the electric field. The energy they gain is dissipated on the front of the tube causing it to light up.

The potential difference is supplied for the plates by a power supply which acts like a large battery.

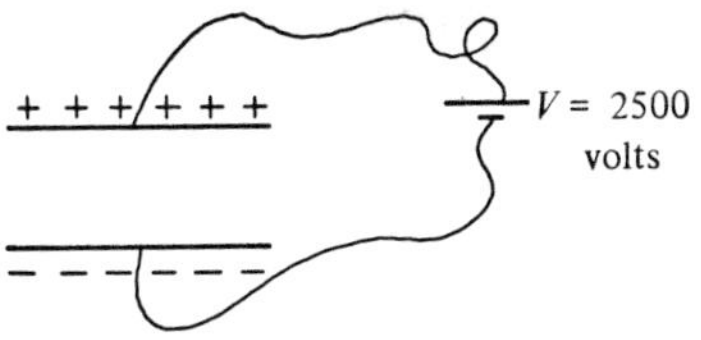

The battery is responsible for the charge on the plates. A device like this with two parallel plates is called a capacitor. The charge on the plate is proportional to the potential difference. What is the proportionality constant for this particular problem?

$C = 1.0 \times 10^{-12}$ farads

$$Q = CV$$

$$\frac{2.5 \times 10^{-8} \text{ coul}}{2.5 \times 10^{2} \text{ volts}} = 1.0 \times 10^{-10} \text{ farads}$$

2.9 The proportionality constant characterizes the parallel plate configuration and is called the capacitance. You will note that typical values of capacitance are quite small in practice.

For a fixed plate area and separation the ratio Q/V is constant. Thus, if you attach a larger battery (bigger V) the amount of charge will be correspondingly higher. From our defining equation of potential difference, if we increase $V_B - V_A$ by attaching a larger battery, what will happen to **E** between the plates?

It will increase.

Problem 3

An insulated conducting sphere of radius 20 cm is given a charge of 5×10^{-8} coul. What is the electric potential at the surface of the sphere?

3.1 The definition of potential difference between two points in an electric field is

$$V_B - V_A = -\int_A^B \mathbf{E} \cdot \Delta \mathbf{s}$$

where the righthand side is in fact the work per unit charge necessary to move charge from point A to point B.

In the problem what is the source of **E**?

The charge on the sphere.

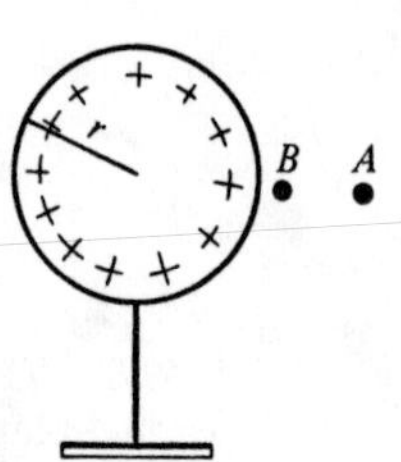

3.2 As in sample problem 1 of this chapter we will choose point A at infinity and call $V_A = 0$. It is V_B we seek.

Indicate on the diagram the electric field configuration in the space around the sphere. Use field lines (arrows).

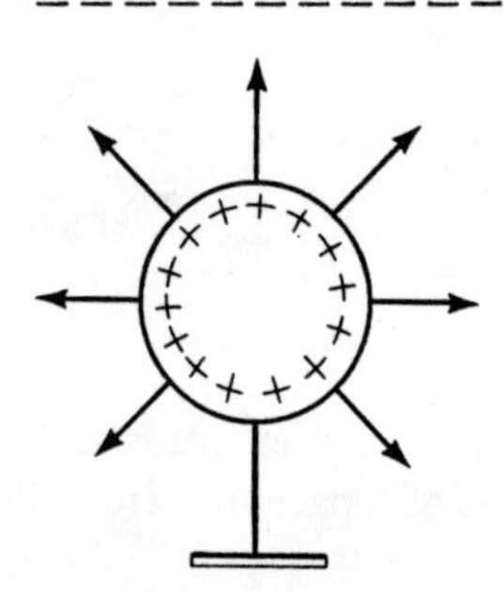

The field lines are all radially directed. Each is perpendicular at the surface of the conductor.

3.3 To use the equation of frame 3.1 we require an expression for **E**. From problem 3 in the chapter on electric fields

$$E = \frac{1}{4\pi\epsilon_0} \frac{q}{R^2} \quad (\text{for } R > r)$$

with a direction as indicated in the diagram of frame 3.2.

This problem is similar to problem 1 in this chapter. Actually this sphere acts like a point charge at the center. We will use the result of problem 1. Refer to frame 1.12 and write the answer in algebraic form.

$V_B =$ ____________________

$$V_B = \frac{1}{4\pi\epsilon_0} \frac{q}{R}$$

3.4 Obtain the numerical result.

$V_B =$ ____________________ volts

$V_B = 2.25 \times 10^3$ volts

$$V_B = \frac{1}{4\pi\epsilon_0} \frac{q}{R}$$

$$V_B = \frac{9 \times 10^9 \times 5 \times 10^{-8}}{2 \times 10^{-1}}$$

$$V_B = 2.25 \times 10^3 \text{ volts}$$

3.5 One last comment is in order. The charge in this problem is about what one might expect to appear on a balloon when you rub it through your hair or on wool trousers. Thus, the balloon has an electric potential of serveral thousand volts. Why doesn't it kill you to hold such a charged object even if you are grounded?

The balloon does not have enough charge to maintain a continuous current flow of sufficient magnitude for injury.

SELF–TEST

1 A popular lecture demonstration is a large metal sphere of radius 30 cm which can be charged to very large electric potential. What is the charge on such a sphere when the electric potential is 50,000 volts? How many electrons would be required to establish the charge?

2 The potential difference between two parallel conducting plates is 50 volts. The plates are separated by 5 cm. An electron placed at rest between the plates will be accelerated. Determine the acceleration.

3 A point charge of 6×10^{-8} coul is located at the origin of a coordinate system. How much work is required to move an electron from $x = 3$ m to $x = 6$ m? What is the potential difference between the two points?

4 Charges $-5q$ and $+2q$ are located at the base of an equilateral triangle. If the magnitude of the charges is 10^{-8} coul, what is the length of each side of the triangle if the electric potential at the apex is -3×10^3 volts?

Answers to Self-Test

1 $q = 1.7 \times 10^{-6}$ coul, about 1×10^{16} electrons

2 1.8×10^{13} m/sec^2

3 $W = +1.44 \times 10^{-17}$ joules
$\Delta V = 90$ volts

4 18 cm on each side

CHAPTER FOUR
Magnetic Fields

If the sample problems and objectives below identify your weak points, go directly to the programmed study section on page 59. If not, try the problems and compare your answers with those that follow. If you can do all the problems easily and if you are familiar with the objectives, you may wish to skip all or part of this chapter. The programmed study section covers techniques and concepts basic to solving the sample problems and fulfilling the objectives in this chapter. A programmed, step-by-step solution of each sample problem begins on page 70. A self-test is included at the end of the chapter.

SAMPLE PROBLEMS AND OBJECTIVES

Problem 1

At the equator the earth has a *N-S* magnetic field of approximately 0.4×10^{-4} weber/m^2. What *W-E* velocity would a proton require in order to maintain an equatorial orbit due to the interaction of the moving proton with the earth's magnetic field? (Use $q = 1.6 \times 10^{-19}$ coul and $m = 1.67 \times 10^{-27}$ kg for the proton. Use $R = 6 \times 10^{6}$ m for the earth.)

Objectives:
1. Discussing work done by forces of the type $\mathbf{F} = q\mathbf{v} \times \mathbf{B}$.
2. Reviewing the dynamics of uniform circular motion.

Problem 2

Show that a magnetic field can be utilized to focus a beam of diverging charged particles. Show further that the motion is helical, and calculate the distance a particle travels down the beam in one period of the helix.

Objectives:
1. Discussing the direction of **F** for various combinations of $\mathbf{v} \times \mathbf{B}$.
2. Decomposition of motion into accelerated and non-accelerated components.

Problem 3

A wire in the shape shown has a current i. Use the Biot-Savart law to calculate the magnetic field **B** at P.

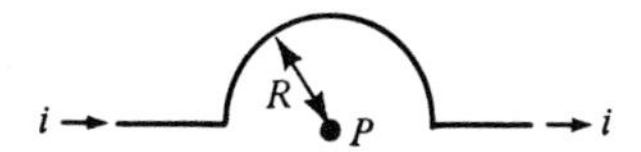

Objectives:
1. Reviewing cross-products.
2. Applying the Biot-Savart law.

Answers to Sample Problems

See page 70 for programmed, step-by-step solutions to these problems.

Problem 1

$v = 15.1 \times 10^4$ m/sec

Problem 2

$z = \frac{2\pi v m}{qB}$ (for charged particles originally diverging from a point)

Problem 3

$B = \frac{\mu_0 i}{4a}$ (into the paper at P)

PROGRAMMED STUDY SECTION

In the chapter on electric fields we began by considering the effect of such fields on test charges, operationally defining the field as $\mathbf{E} = \mathbf{F}/q_0$. Following this we considered the source of electric fields as due to fixed charges and charge distributions using Gauss's law to determine the resulting electric fields. Similarly this programmed study section will emphasize operationally defining a magnetic field and will consider the source of magnetic fields.

1 To clarify the operational definition developed later, let us start with the more familiar case of the electric field. Ignoring other types of interaction we say that the region of space in which we place a positive test charge q_0 has an electric field **E** if q_0 experiences a force **F** due to the electric field.

We idealize the "probing" in the sketch by attaching a test charge to a spring scale to see if an electric field exits. The scale registers some reading. What is the direction of E at the location of q_0?

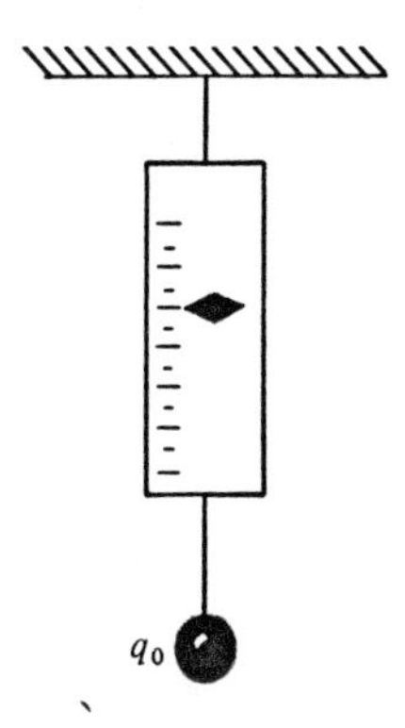

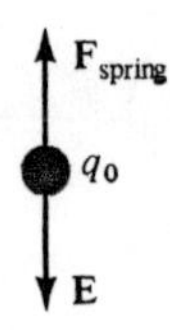

The spring force is counteracting the electric force

$$\mathbf{F}_{elec} = -q_0\mathbf{E}$$

so **E** must be opposite to the spring force.

2 We could thus use this little probe to find both the magnitude and direction of the electric field **E** at various points. For magnetic fields we will use a probe which consists of moving charges.

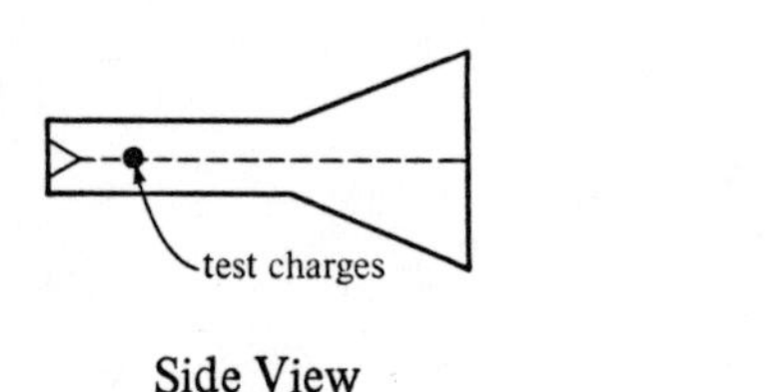

Side View

Front View

To make the discussion more graphic we will use a special TV tube which contains a stream of test charges. If the test charges do not interact with a magnetic field then the spot on the front of the screen (which implies the motion of the charges) will remain stationary. Similarly, our spring scale would indicate zero in the absence of an electric field.

The amount of deflection is a measure of the strength of interaction. The pictures below depict the spot position under varying conditions. **B** is the same in each case.

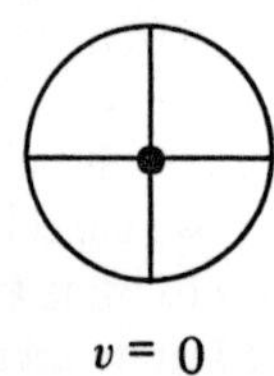

$v = 0$

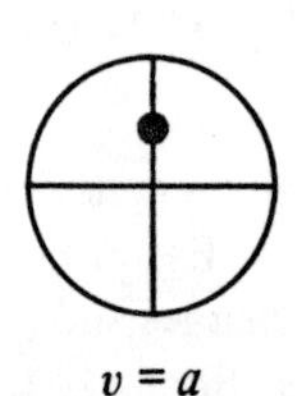

$v = a$

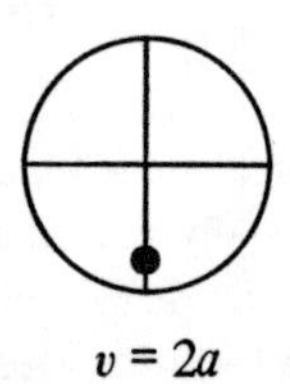

$v = 2a$

How does the strength of interaction of the test charges depend on the speed of the test particles?

The strength of the interaction is proportional to the speed of the test particles.

3 Show by arrows on the tubes the relative directions of **v** and **B** for the interaction evidence shown below.

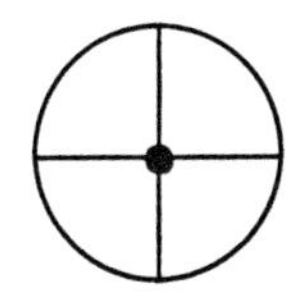

$\mathbf{B} \parallel \mathbf{v}$

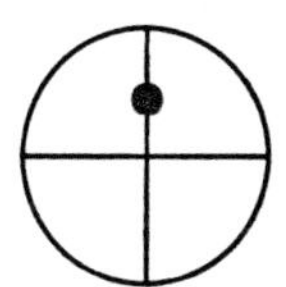

B makes an angle of 30° with **v**.

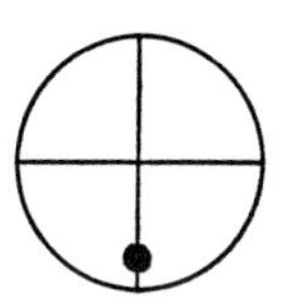

$\mathbf{B} \perp \mathbf{v}$

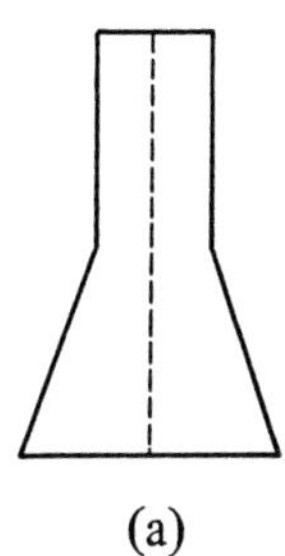

(a)

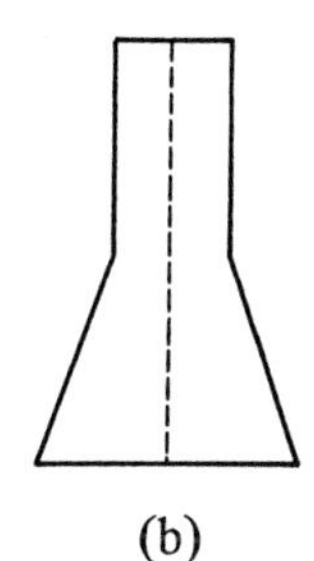

(b)

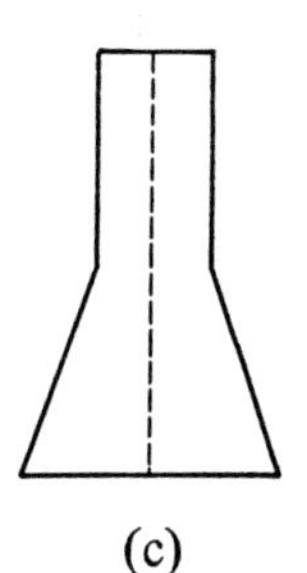

(c)

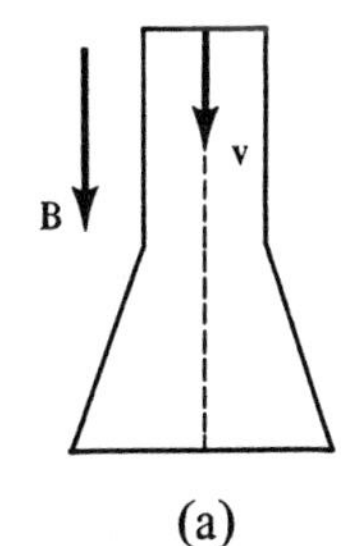

(a)

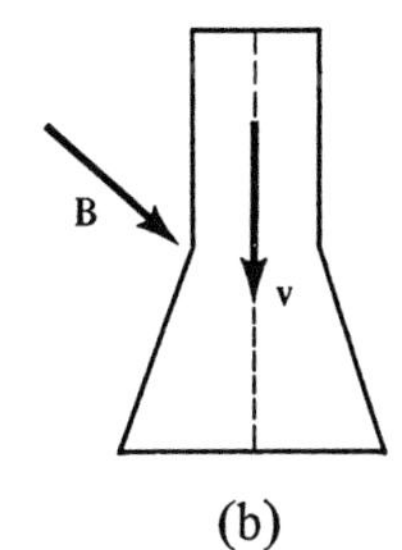

(b)

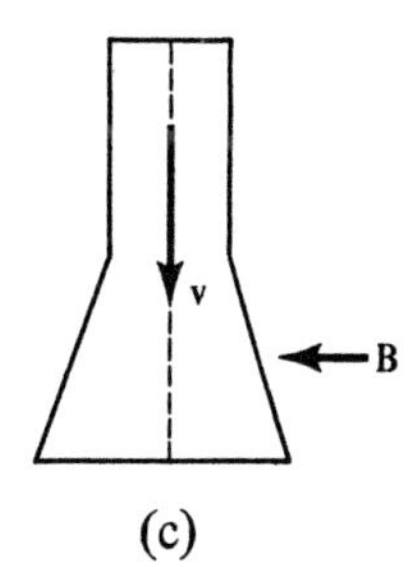

(c)

4 We see from the previous frame that the interaction is strongest when $\mathbf{v} \perp \mathbf{B}$. Using the previous answer, complete the statements below.

(a) The angle between **v** and **B** = ______________ .

(b) The angle between **v** and **B** = ______________ .

(c) The angle between **v** and **B** = ______________ .

(a) 0° (parallel); (b) 30°; (c) 90° (perpendicular)

5 For each case of frame 4, compute $\sin\theta$ where θ is the angle between the direction of **v** and the direction of **B**.

(a) $\sin\theta =$ __________

(b) $\sin\theta =$ __________

(c) $\sin\theta =$ __________

- - - - - - - - - - - - - - - - - - - -

(a) 0; (b) 0.5; (c) 1.0 (Note that $\sin\theta$ is largest for the case in which the interaction is strongest.)

6 A description of the interaction between moving test charges and magnetic fields must include

(a) the speed of the test charge

(b) the relative orientation of **v** and **B**

(c) ______________________________

Add another requirement (c) consistent with the diagrams below. (**B** is ⊥ to **v** in each case.)

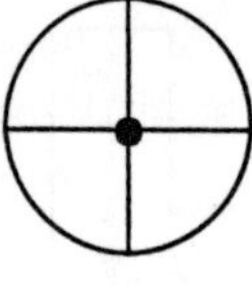

(a) $B = 0$

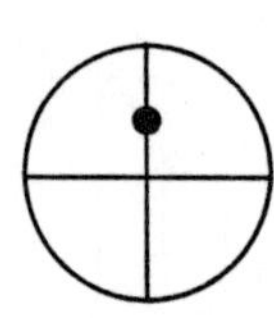

(b) $\mathbf{B} = \mathbf{B}_0$

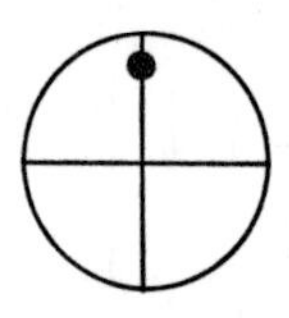

(c) $\mathbf{B} = 2\mathbf{B}_0$

- - - - - - - - - - - - - - - - - - - -

(c) the magnitude of the magnetic field **B**

7 Finally the interaction also depends on the magnitude of the charge. All elements of the description are included in a special notation.

$$\mathbf{F} = q\mathbf{v} \times \mathbf{B}$$

$F = qvB \sin\theta$ (where θ is the angle between **v** and **B**)

This notation is called the cross-product.

We must now determine the direction of **F** itself. This is given by the right-hand rule. Hold the fingers of your right hand in the direction of **v**, close the hand normally so the fingers move toward **B**, and the thumb (erect) gives the direction of **F**.

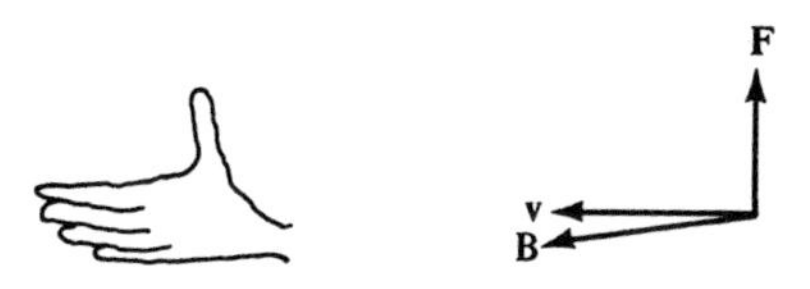

Try the examples below. Indicate the direction of **F** in each case.

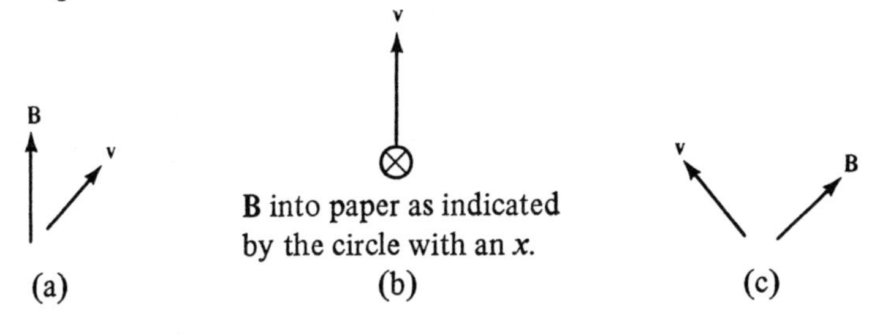

(a) **F** ⊥ **v** and **B** pointing out of paper; (b) **F** in plane of paper ⊥ to **v** and **B** and pointing left; (c) **F** ⊥ **v** and **B** pointing into the paper

8 The prescription then is to probe with a test charge q which has a velocity **v** and see if it experiences a force **F** described by $\mathbf{F} = q\mathbf{v} \times \mathbf{B}$.

To test your understanding of the right-hand rule answer the question below.

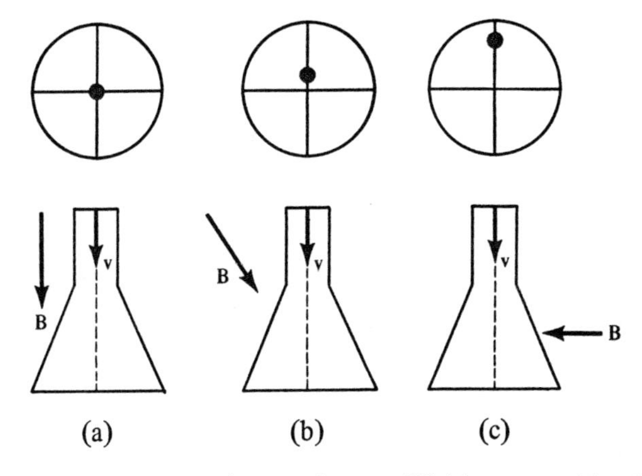

v and **B** have the same magnitude in each case. Which spot position is *incorrect?*

(c) It should be like this:

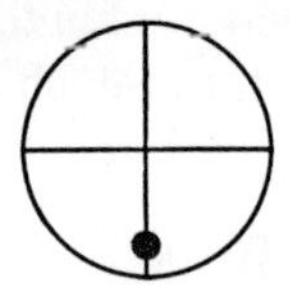

9 One final remark concerning the force of interaction between a moving test charge and a magnetic field. The vectors **v** and **B** can be thought of as lying in the same plane. Thus, **F** is always *perpendicular* to this plane.

10 We now begin to review the sources of magnetic fields by discussing the physical laws and their meaning. Ampere's law is written:

$$\Sigma \mathbf{B} \cdot \Delta \ell = \mu_0 \Sigma i$$

where the lefthand side is the sum of many terms.

$$\Sigma \mathbf{B} \cdot \Delta \ell = B_1 \Delta \ell_1 \cos \theta_1 + B_2 \Delta \ell \cos \theta_2 + \cdots$$

B_1 is essentially constant over the distance $\Delta \ell_1$, and θ_1 is the angle between B_1 and $\Delta \ell_1$. We will discuss the right side shortly.

One requirement of Ampere's law is that the sum form a closed path (i.e., all the $\Delta \ell$'s must add up to a closed path). In a manner similar to Gauss's law the closed path is imaginary and can be selected for convenience. You will recall that Gaussian surfaces were not necessarily actual physical surfaces.

Consider the closed path *abcd* shown. Diagrammatically indicate a $\Delta \ell$ as implied in Ampere's law.

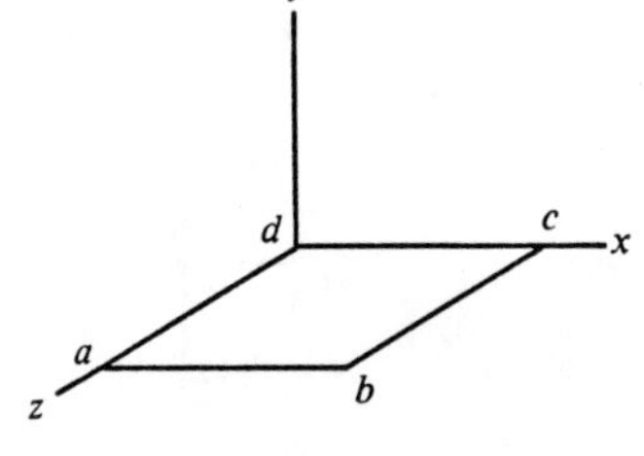

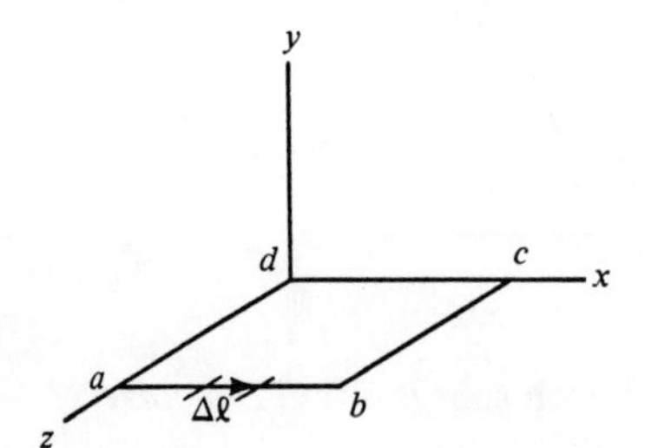

You may have placed yours in a different position. Notice the $\Delta \ell$ shown implies going around the closed path from *a* to *b* to *c* to *d* and back to *a*.

11 Here we indicate a constant magnetic field **B**. We show only one arrow, but it is understood that the magnetic field is everywhere the same and has only components in the y direction.

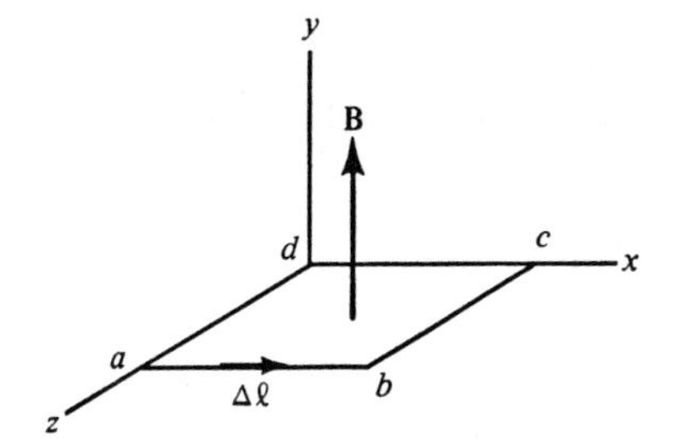

Ampere's law requires

$$\sum \mathbf{B}\cdot\Delta\ell = \sum_a^b \mathbf{B}\cdot\Delta\ell + \sum_b^c \mathbf{B}\cdot\Delta\ell + \sum_c^d \mathbf{B}\cdot\Delta\ell + \sum_d^a \mathbf{B}\cdot\Delta\ell$$

where we are trying to show that the path must be closed. For this particular case each term is zero. Why is each term zero?

B is always $\perp$ to $\Delta\ell$.

$\mathbf{B}\cdot\Delta\ell = B\Delta S \cos\theta$
$\theta = 90^\circ, \cos\theta = 0$

12 Here is a slightly differnt case. The magnetic field has components only in the $-z$ direction and is everywhere the same.

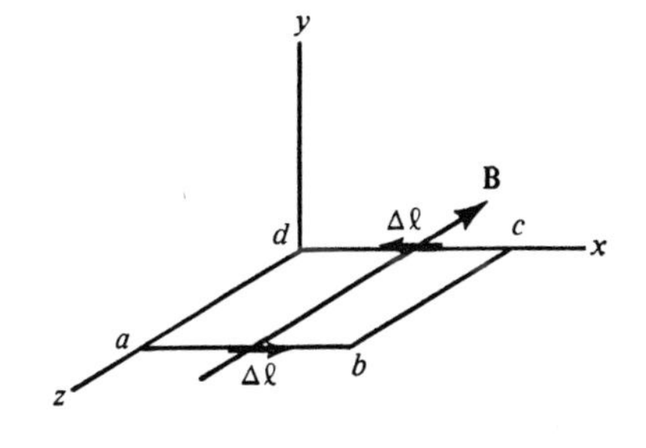

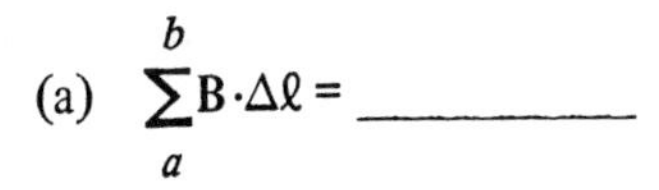

(a) $\sum_a^b \mathbf{B}\cdot\Delta\ell =$ ________

(b) $\sum_c^d \mathbf{B}\cdot\Delta\ell =$ ________

Zero in both cases because **B** is always $\perp$ $\Delta\ell$.

13 Remember that we must consider a complete path. So far for this new case shown below the answer is zero since $\cos 0^\circ = 1$ and B is constant.

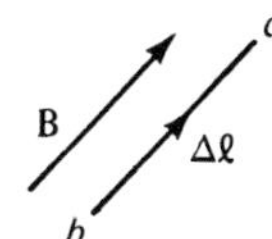

$$\sum_b^c \mathbf{B}\cdot\Delta\ell = B \sum_b^c \Delta\ell$$

Describe geometrically the term $\sum_b^c \Delta\ell$.

This is the distance from b to c.

14 For three-quarters of the way around the closed path we have

$$\sum \mathbf{B}\cdot\Delta\ell = \underbrace{\sum_a^b \mathbf{B}\cdot\Delta\ell}_{=0} + \underbrace{\sum_b^c \mathbf{B}\cdot\Delta\ell}_{=B\ell} + \underbrace{\sum_c^d \mathbf{B}\cdot\Delta\ell}_{=0} + \sum_d^a \mathbf{B}\cdot\Delta\ell$$

where ℓ is the distance from c to b. What is the value of the last term on the right? Be careful!

$-B\ell$

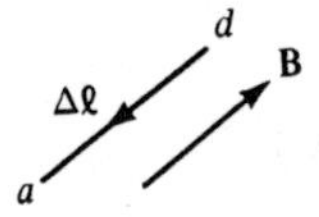

$\mathbf{B}\cdot\Delta\ell = -B\Delta\ell$ since $\cos 180° = -1$. Thus, $\sum_d^a \mathbf{B}\cdot\Delta\ell = -B\ell$.

ℓ is the distance from d to a as it is from c to b.

15 Adding everything up we have

$$\Sigma\mathbf{B}\cdot\Delta\ell = 0 + B\ell + 0 - B\ell = 0$$

for the case depicted in frame 12. Now Ampere's law relates $\Sigma\mathbf{B}\cdot\Delta\ell$ around a *closed* path to the net electric current *enclosed* by that *path.* In what way is this reminiscent of Gauss's law for electric fields?

Gauss's law relates the flux of **E** through a closed *surface* to net charge *enclosed by the surface.*

16 What must be the net current enclosed by the path *abcd* in both frames 11 and 12?

Zero (since $\Sigma\mathbf{B}\cdot\Delta\ell = 0$ in both cases)

17 Thus, we see the significance of the righthand side of Ampere's law

$$\Sigma\mathbf{B}\cdot\Delta\ell = \mu_0 \Sigma i$$

It is the net current enclosed by some closed path. In the two examples given $\mathbf{B} \neq 0$, but the net current enclosed was zero, therefore $\Sigma\mathbf{B}\cdot\Delta\ell = 0$.

Let us take one final look at this law to make sure you know what the terms mean. The coil shown has a current flowing in the wire. Show by an arrow the magnetic field produced by such a coil. (We have not discussed this, but most people know the answer.)

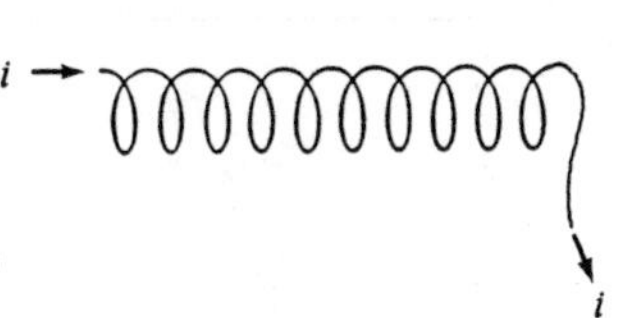

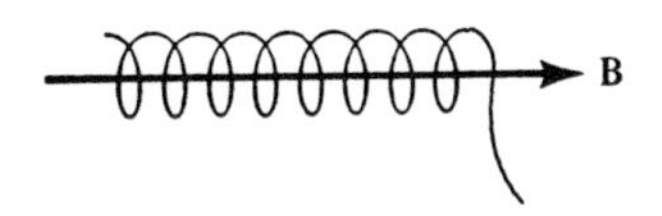

The field is fairly constant inside the coil. Curve the fingers in the direction of i and the thumb gives the direction of **B**.

18 Now let us look at a cross-section of the coil. We show a slice of the coil with electric current coming out of the top of the coil and into the bottom of the adjacent coil. Also shown are three close paths marked 1, 2, and 3. Which of the closed paths enclose net currents?

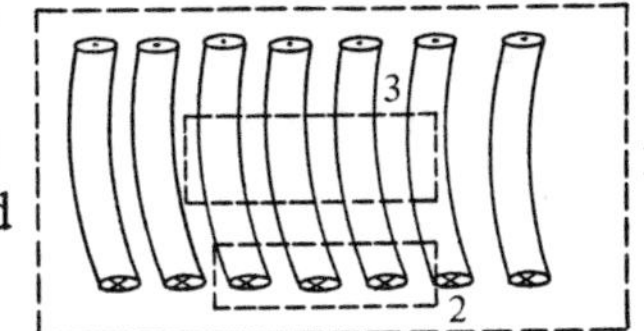

2

19 Does Ampere's law require that **B** around the closed path be everywhere zero for path 3? (Check the answer to frame 17 before you give a hasty reply.)

No. **B** is not zero everywhere along the closed path 3. As a matter of fact, it is constant everywhere along the path. Ampere's law requires that $\Sigma \mathbf{B} \cdot \Delta \ell = 0$ when the enclosed current is zero.

20 For the final portion of the programmed study frames we will consider another law relating magnetic fields to their source. The symbols are a little tricky, but they will hopefully be clear if we relate them to pictures.

We show a wire carrying a fixed current i. We wish to know the magnetic field **B** at P due to this wire. The Biot-Savart law considers the field contribution $\Delta \mathbf{B}$ at P of a small current element $\Delta \ell$. The law reads

$$\Delta \mathbf{B} = \frac{\mu_0 i}{4\pi} \frac{\Delta \ell \times \hat{\mathbf{e}}_r}{r^2}$$

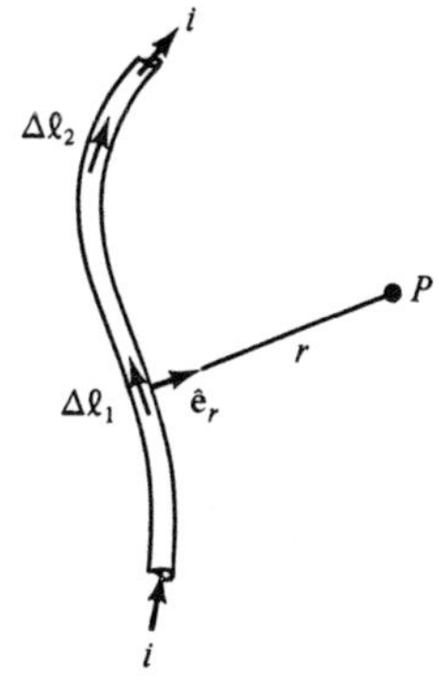

The vector $\Delta \ell$ is clearly associated with the direction of the current. The unit vector $\hat{\mathbf{e}}_r$ has magnitude 1 and points from $\Delta \ell$ to P. Show on the sketch the unit vector properly associated with current element $\Delta \ell_2$.

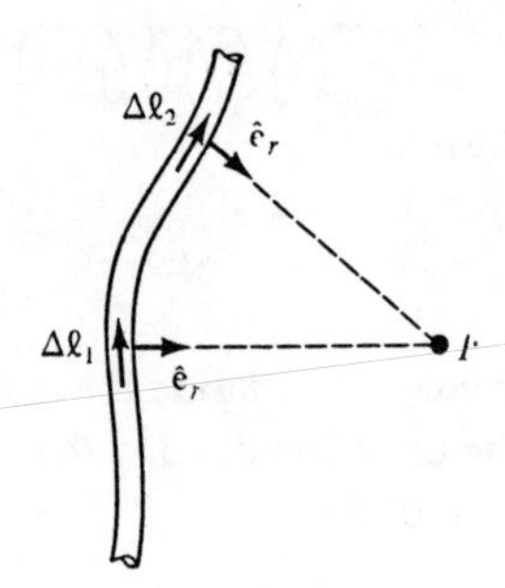

The unit vector points from $\Delta \ell$ to the point P in question.

21 Earlier in this chapter the cross-product of **v** and **B** was discussed. In the Biot-Savart law we again encounter the cross-product. We can write the law as

$$\Delta \mathbf{B} = \frac{\mu_0 i}{4\pi r^2} \Delta \ell \times \hat{\mathbf{e}}_r$$

How must $\Delta \mathbf{B}$ be spatially related to $\Delta \ell$ and $\hat{\mathbf{e}}_r$?

Perpendicular to both (We can think of $\Delta \ell$ and $\hat{\mathbf{e}}_r$ as being in a plane.)

22 The direction of $\Delta \mathbf{B}$ at P due to the current element $\Delta \ell$ is given by the right-hand rule for $\Delta \ell \times \hat{\mathbf{e}}_r$. Both $\Delta \ell$ and $\hat{\mathbf{e}}_r$ are in the plane of the paper. What is the direction of $\Delta \mathbf{B}$ at P?

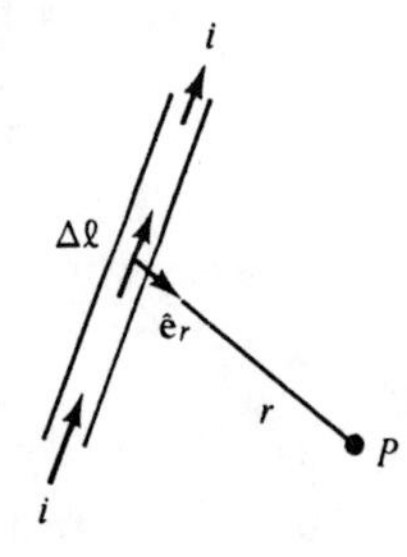

Perpendicular to the paper and pointing into the paper at P.

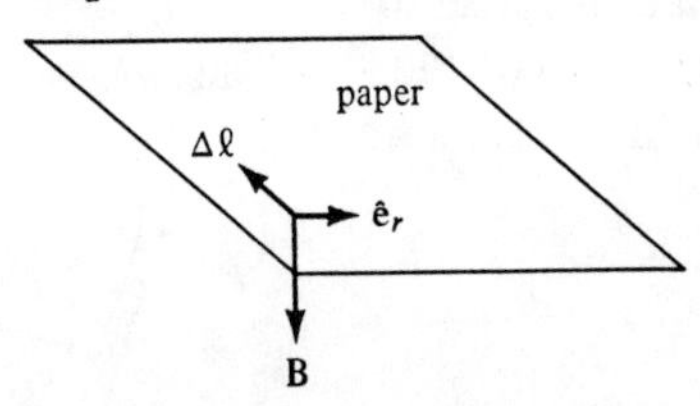

23 The variable r in the Biot-Savart law is the distance from a particular current element $\Delta \ell$ to the point at which one wishes to determine the magnetic field $\Delta \mathbf{B}$. In general r is different for each current element $\Delta \ell$.

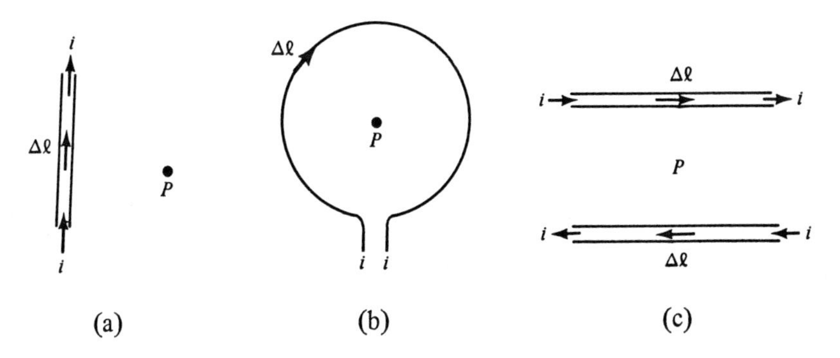

In each case above, identify $\hat{\mathbf{e}}_r$ and r for the current element $\Delta \ell$.

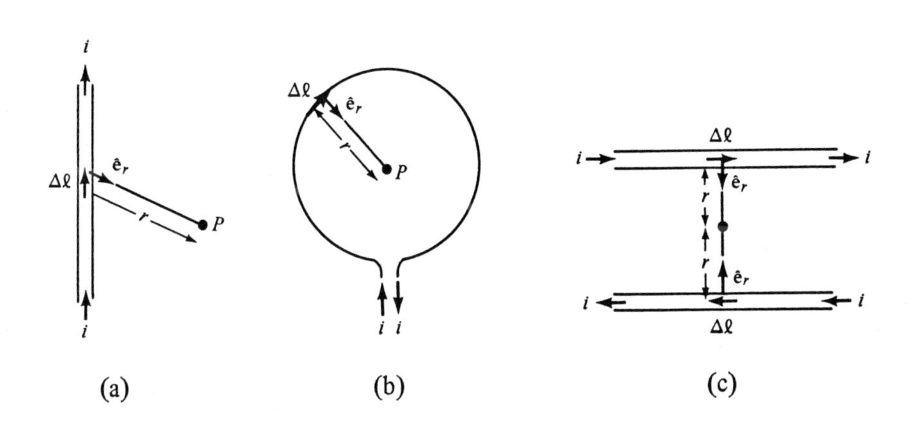

24 One final remark about the meaning of this law. By looking at the previous answer, comment on the direction of $\Delta \ell$ compared to the direction of current.

By convention $\Delta \ell$ and i are in the same direction.

SOLUTIONS TO SAMPLE PROBLEMS

Problem 1

At the equator the earth has a *N-S* magnetic field of approximately 0.4×10^{-4} weber/m^2. What *W-E* velocity would a proton require in order to maintain an equatorial orbit due to the interaction of the moving proton with the earth's magnetic field? (Use $q = 1.6 \times 10^{-19}$ coul and $m = 1.67 \times 10^{-27}$ kg for the proton. Use $R = 6 \times 10^6$ m for the earth.)

1.1 Certain forces in physics often result in particular consequences. For example, we know that the force $F = -kx$ resulted in periodic motion. Let us begin this problem by discussing the consequences of the force

$$\mathbf{F} = q\mathbf{v} \times \mathbf{B}$$

What is the spatial relationship between **F** and **v** as discussed in the programmed study section?

F and **v** are always perpendicular.

1.2 We know that $\mathbf{v} = \Delta\mathbf{s}/\Delta t$. What is the spatial relationship between $\Delta\mathbf{s}$ and **v**?

$\Delta\mathbf{s}$ and **v** are always parallel.

1.3 For a force law

$$\mathbf{F} = q\mathbf{v} \times \mathbf{B}$$

what is the spatial relationship between **F** and Δs where Δs is the displacement of q under the influence of **F**?

F is always perpendicular to $\Delta\mathbf{s}$.

$\mathbf{F} \perp \mathbf{v}$ (frame 1.1)
$\mathbf{v} \parallel \Delta\mathbf{s}$ (frame 1.2)

1.4 Taking this last result, we can ask how much work is done by a force of the type $\mathbf{F} = q\mathbf{v} \times \mathbf{B}$ when acting on a charge q. For example

$$\Delta W = \mathbf{F} \cdot \Delta\mathbf{s} = F\Delta s \cos\theta$$

What is the value of ΔW?

- - - - - - - - - - - - - - - - - -

$\Delta W = 0$ always since $\cos\theta = 0$ when $\theta = 90°$.

1.5 In mechanics we use the theorem

$$\text{Work} = \text{change in kinetic energy}$$
$$W = \Delta K$$

What does the theorem predict for the change in speed of a charge q which is acted upon by a force $\mathbf{F} = q\mathbf{v} \times \mathbf{B}$?

- - - - - - - - - - - - - - - - - -

The speed will not change. $W = 0$ $\Delta K = 0$

$\frac{1}{2}mv^2$ = constant for the charge q

1.6 Let us review the frames thus far.

- A force acts on a particle of mass m, with velocity $\mathbf{v}$ and charge q. The force is $\mathbf{F} = q\mathbf{v} \times \mathbf{B}$.
- The force can do no work on the particle so the particle has a constant speed.
- The force is always perpendicular to the displacement.

Can such a situation be true of an accelerated particle?

- - - - - - - - - - - - - - - - - -

Yes. $\mathbf{a} = \mathbf{F}/m$ There is indeed a force acting on the particle (Newton's second law).

1.7 How can a particle be accelerated even though its speed cannot change?

- - - - - - - - - - - - - - - - - -

The direction of the velocity can change. Speed is the magnitude of the velocity.

1.8 If we further stipulate that $\mathbf{v}$ has no component in the direction of $\mathbf{B}$ then we have:

- The force is perpendicular to the displacement.
- The particle is accelerated.
- The particle has constant speed.

What kind of motion is characterized by these facts?

Uniform circular motion

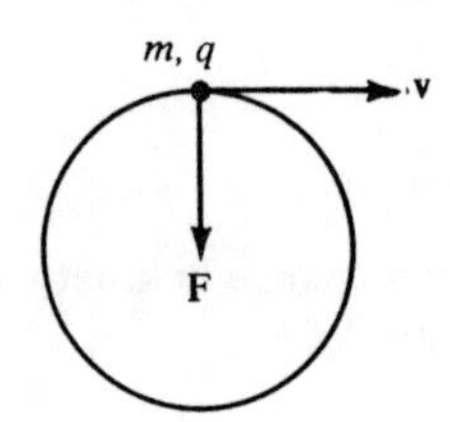

B is considered as coming out of the paper.

1.9 The special form of Newton's second law for uniform circular motion is

$F =$ ____________

$$F = m\frac{v^2}{R}$$

1.10 It is, of course, necessary that **F** in the previous answer be centripetal (i.e., towards the center). Consider the cases below and indicate the actual direction of $\mathbf{F} = q\mathbf{v} \times \mathbf{B}$. The charge is considered positive. **B** is into the paper when x's are used and out of the paper for dots. This is a standard notation.

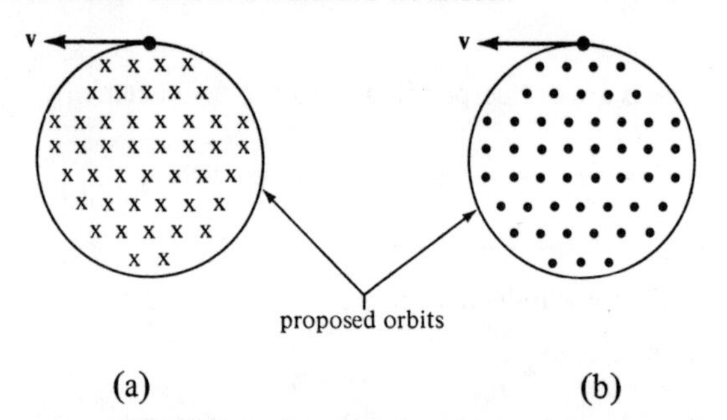

Use arrows to indicate **F** acting on each charge.

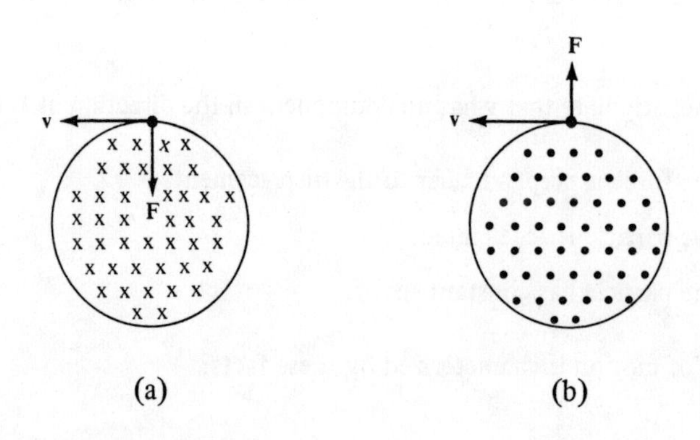

1.11 Of the two proposed orbits only case (a) is consistent with the directions of **v** and **B** as shown. What is the magnitude of **F** in case (a)?

$F = qvB$

v and **B** are perpendicular, therefore the angle between these two vectors is 90°.

$$\mathbf{F} = q\mathbf{v} \times \mathbf{B}$$
$$F = qvB \sin 90°$$
$$F = qvB$$

1.12 Using answer frame 1.9 and 1.11 we have

$$qvB = \frac{mv^2}{R}$$

as the working equation for this problem. We see the speed of a proton necessary to maintain a circular orbit in the earth's magnetic field. Indeed, at the equator the magnetic field is perpendicular to the proposed path for the proton. Use the following to determine v.

$q = 1.6 \times 10^{-19}$ coul $\qquad m = 1.67 \times 10^{-27}$ kg
$R = 6 \times 10^6$ m $\qquad B = 0.4 \times 10^{-4}$ weber/m²

$v = 15.1 \times 10$ m/sec

$$v = \sqrt{\frac{1.6 \times 10^{-19} \times 4 \times 10^{-5} \times 6 \times 10^6}{1.67 \times 10^{-27}}}$$
$$v = \sqrt{230 \times 10^8}$$
$$v = 15.1 \times 10^4 \text{ m/sec}$$

Problem 2

Show that a magnetic field can be utilized to focus a beam of diverging charged particles. Show further that the motion is helical, and calculate the distance a particle travels down the beam in one period of the helix.

2.1 To focus means to bring the charged particles together at a point. We will consider one particle and generalize from the result. A source of charged particles is situated at the origin. One of them leaves the origin as shown with the following velocity components.

$$v_x = v \sin\theta$$
$$v_y = 0$$
$$v_z = v \cos\theta$$

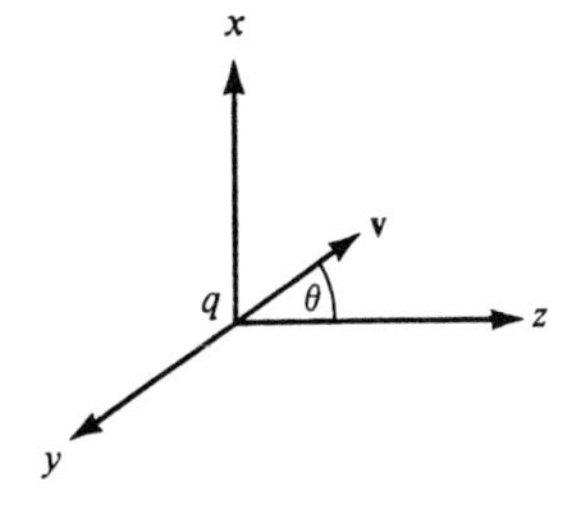

It is to be understood that the magnetic field points in the z direction only. Which of the velocity components on page 73 will be unaffected by the force of the general form $\mathbf{F} = q\mathbf{v} \times \mathbf{B}$?

$v_z = v \cos \theta$ (v_z is parallel to **B** and therefore no change will occur in v_z.)

2.2 The component v_x will be affected. What is the magnitude of the force on q due to the force of general form $\mathbf{F} = q\mathbf{v} \times \mathbf{B}$?

$F = qv \times B$ (v_x and **B** are perpendicular.)

2.3 In describing projectile motion near the surface of the earth one talks of an unaffected horizontal velocity and an accelerated vertical velocity.

For the moment we will concentrate on what happens to v_x since v_z is not going to change. Imagine looking back along the z axis at the positively charged particle. We show v_x and it is understood that **B** points toward you (along the z axis). Draw a vector representing **F** due to the interaction of q, v_x, and **B**.

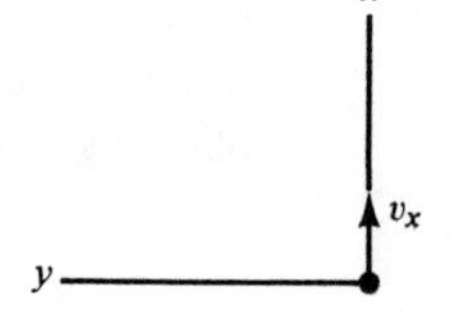

F is in the $-y$ direction.

2.4 Indicate by a dotted line the resulting motion of q due to a force of this type. Remember that **v** and **F** are always perpendicular for forces of the type $\mathbf{F} = q\mathbf{v} \times \mathbf{B}$.

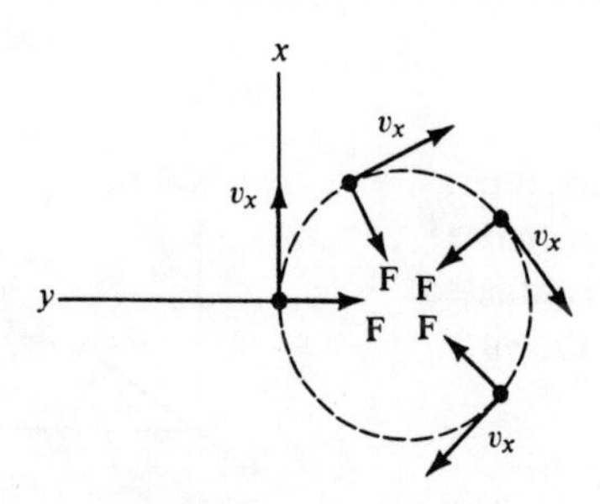

The motion of the particle in the x-y plane is uniform circular motion. Note that **F** is a centripetal force.

2.5 Use Newton's second law to obtain an expression for the radius of the orbit sketched in the previous answer. Express R in terms of v_x, m, q, and B.

$R = \frac{mv_x}{qB}$

$F = m\frac{v^2}{R}$ (for uniform circular motion)

$F = qvB$ (in this case)

$qvB = m\frac{v_x{}^2}{R}$

2.6 If we multiply R by 2π we have the distance D traveled during one revolution.

$$D = 2\pi R = 2\pi\frac{mv_x}{qB}$$

Is the speed of q the same through the entire orbit?

Yes. (This is a characteristic of $\mathbf{F} = q\mathbf{v} \times \mathbf{B}$.

2.7 The period T of the circular orbit is the time required for one revolution.

$2\pi\frac{mv_x}{qB}$ = distance traveled during one orbit

v_x = speed of q in orbit

T = ____________

$T = 2\pi\frac{m}{qB}$

$T = \frac{\text{distance}}{\text{speed}}$

Note that the answer is independent of v_x.

2.8 After a time T the particle will again be somewhere along the z axis. Looking back at frame 2.1, which velocity component is responsible for the change in the z location of the charge q?

---

v_z

2.9 We see that while v_x is making the particle revolve in a circle, v_z is translating the particle along the z axis.

At $t = T = 2\pi\frac{m}{qB}$, z = ____________

$$z = v_z 2\pi \frac{m}{qB}$$

2.10 Here we show the distance z traveled by the particle while it was completing a revolution. The combination is a helix.

$$v_z = v \cos\theta$$
$$z = v \cos\theta \; 2\pi \frac{m}{qB}$$

We will invoke the approximation that $\cos\theta$ is close to 1 for small angles.

$$z = \frac{2\pi v m}{qB}$$

The above equation can be applied to all three charged particles shown below. Each would come together again (i.e., focus) at $z = 2\pi v m/qB$. Clearly an adjustment of the magnetic field **B** allows z to change.

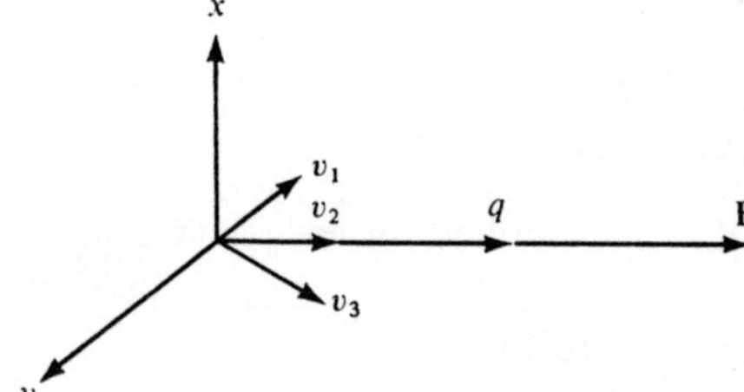

Problem 3

A wire in the shape shown has a current i. Use the Biot-Savart law to calculate the magnetic field **B** at P.

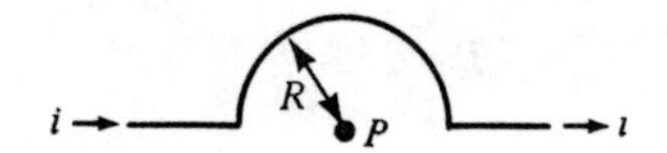

3.1 Using the Biot-Savart law to find **B** at P as shown to the right we must add the $\Delta\mathbf{B}$'s contributed by all elements of current $\Delta\ell$. We show three representative $\Delta\ell$'s. To apply

$$\Delta\mathbf{B} = \frac{\mu_0 i}{4\pi} \frac{\Delta\ell \times \hat{\mathbf{e}}_r}{r^2}$$

for each of the elements shown, draw the appropriate vector $\hat{\mathbf{e}}_r$ for *each* of the individual current elements.

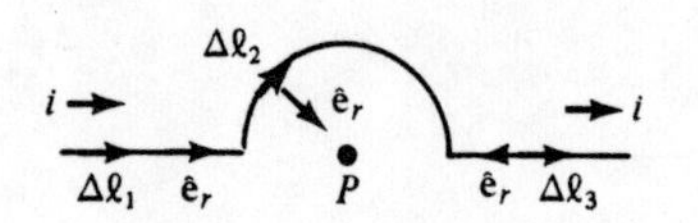

Note that $\hat{\mathbf{e}}_r$ points from the current element to the point at which **B** is to be determined.

3.2 The cross product of two vectors has a magnitude which is determined as follows:

$$\mathbf{A} \times \mathbf{B} = AB \sin \theta$$

where θ is the angle between **A** and **B**. What is the particular value of

(a) $\Delta \ell_1 \times \hat{\mathbf{e}}_r =$ ____________________

(b) $\Delta \ell_3 \times \hat{\mathbf{e}}_r =$ ____________________

Both are zero.

$\ell_1 \times \hat{\mathbf{e}}_r = \ell_1 \hat{\mathbf{e}}_r \sin 0°$ (in magnitude)
$\ell_3 \times \hat{\mathbf{e}}_r = \ell_3 \hat{\mathbf{e}}_r \sin 180°$ (in magnitude)
$\sin 0° = \sin 180° = 0$

$\Delta \ell_1$, $\hat{\mathbf{e}}_r$, $\theta = 0°$ $\Delta \ell_3$, $\hat{\mathbf{e}}_r$, $\theta = 180°$

3.3 In view of the previous answer we need only consider the curved portion of the wire. The straight segments do not contribute to the field **B** at P.

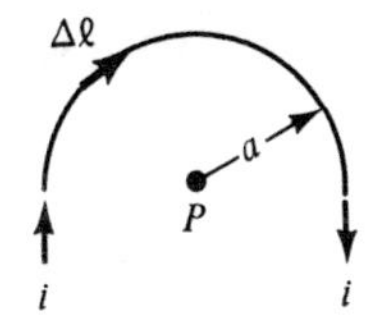

$$\Delta \mathbf{B} = \frac{\mu_0 i}{4\pi} \frac{\Delta \ell \times \hat{\mathbf{e}}_r}{r^2}$$

Indicate on the diagram $\Delta \mathbf{B}$, $\hat{\mathbf{e}}_r$, and r for the $\Delta \ell$ shown.

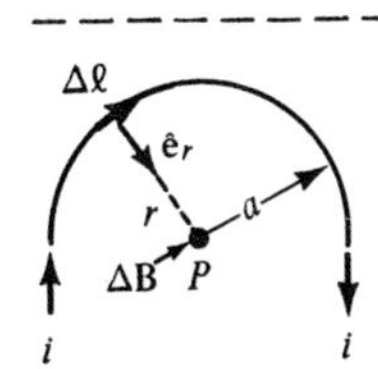

$\Delta \mathbf{B}$ is into the paper at P. (Note: $r = a$.)

3.4 The total field at P will be the sum of all such current elements. Because of symmetry, the angle θ between $\Delta \ell$ and $\hat{\mathbf{e}}_r$ will be __________ degrees for all current elements on the semicircle.

90° (See answer frame 3.3 as one example.)

3.5 Will the direction of $\Delta \mathbf{B}$ at P be the same for all current elements $\Delta \ell$ along the wire?

Yes. Therefore $\mathbf{B}_{\text{Total}}$ will point into the paper at P.

3.6 We know the direction of **B**, so we need only determine the magnitude.

$$B = \Delta B_1 + \Delta B_2 + \cdots$$

Here we indicate that **B** is the sum of all the Δ**B**'s. Write out the expression for the magnitude ΔB using the picture and information of frame 3.3.

$$\Delta \mathrm{B} = \frac{\mu_0 i}{4\pi a^2} \Delta \ell$$

$\Delta \ell \times \hat{e}_r = \Delta \ell \hat{e}_r \sin 90^\circ$ (in magnitude)
$\Delta \ell \times \hat{e}_r = \Delta \ell$ (in magnitude)
$\hat{e}_r$ has a magnitude 1.

3.7 We see now

$$B_T = \Sigma \Delta B = \Sigma \frac{\mu_0 i}{4\pi a^2} \Delta \ell$$

$$B_T = \frac{\mu_0 i}{4\pi a^2} \Sigma \Delta \ell$$

What is $\Sigma \Delta \ell$ for the picture of frame 3.3?

The length of the wire ($\Sigma \Delta \ell = \pi a$, one-half the circumference of a circle of radius a)

3.8 Using the information of the previous frame

$B_T =$ ________

into the paper at point P.

$$B_T = \frac{\mu_0 i}{4a}$$

SELF–TEST

1 Use the Biot-Savart law to calculate the magnetic field at the center of a circular loop of radius 25 cm which has a current of 5 amp.

2 Two long wires as shown have the same current of 2 amp. Use Ampere's law to calculate the magnetic field at P. The distance $d = 2$ m and $a = 6$ m. Indicate direction as well as magnitude.

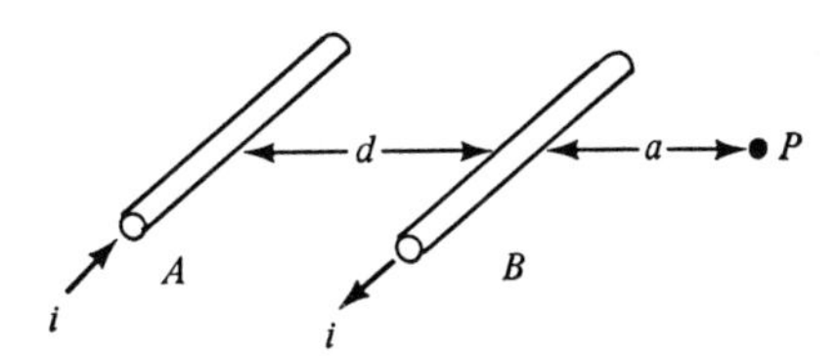

3 What is the orbital radius of a proton which is moving in a circle at right angles to a magnetic field of 1×10^2 weber/m^2? The proton was accelerated through an electric field of 30×10^{-3} nt/coul prior to entering the magnetic field.

Answers to Self-Test

1 1.26×10^{-5} weber/m^2

2 8×10^{-7} weber/m^2

3 $r = 140$ m

CHAPTER FIVE
Faraday's Law of Induction

If the sample problems and objectives below identify your weak points, go directly to the programmed study section on page 81. If not, try the problems and compare your answers with those that follow. If you can do all the problems easily and if you are familiar with the objectives, you may wish to skip all or part of this chapter. The programmed study section covers techniques and concepts basic to solving the sample problems and fulfilling the objectives in this chapter. A programmed, step-by-step solution of each sample problem begins on page 87. A self-test is included at the end of the chapter.

SAMPLE PROBLEMS AND OBJECTIVES

Problem 1

A search coil is a small circular coil which can be used to measure magnetic fields.

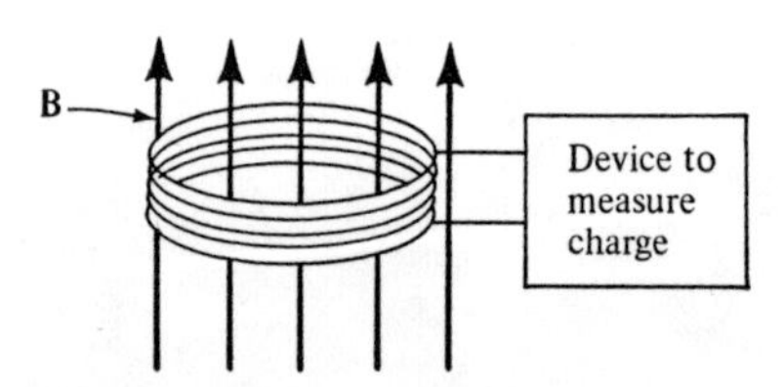

The instrument is aligned as shown with the plane of the coil perpendicular to the unknown **B** and then quickly flipped 90°. Another instrument which measures charge is attached.

Show that Q is proportional to **B** and numerically determine **B** for a search coil of the following description.

$$N = 50 \text{ turns}$$
$$A = 4 \text{ cm}^2$$
$$R = 25\ \Omega$$

where the charge measured during a flip is 4×10^{-5} coul.

Objectives:
1. Discussing magnetic flux and flux changes.
2. Discussing current in terms of charge flow.
3. Numerical example of an actual experimental tool.

Problem 2

The wire W of length $\ell = 0.2$ m moves to the right with constant velocity $v = 0.3$ m/sec. If the resistance of the loop is approximately fixed at 10 Ω and $B = 3.0$ weber/m², what is the magnitude and direction of the induced current?

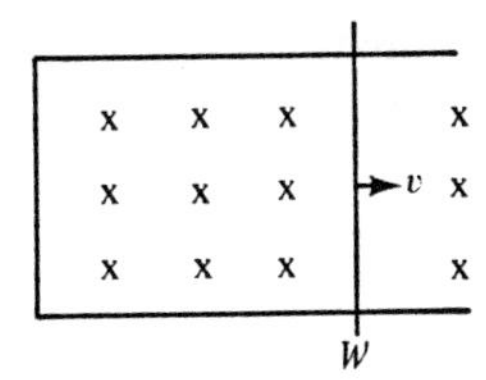

Objectives:
1. Discussing magnetic flux and flux changes with time.
2. Interpreting induced emf's in terms of magnetic forces on moving charges.

Problem 3

A coil in the form of a long solenoid has 1000 turns with a cross-sectional area of 0.1 m² A separate coil of similar cross-section is wound over the primary coil. An emf of 6 volts is induced in the secondary coil if the current in the primary changes by 0.5 ampere in 0.02 sec. How many turns are used for the secondary?

Objectives:
1. Discussion of transformers.
2. Review of the magnetic field of a solenoid.

Answers to Sample Problems

See page 87 for programmed, step-by-step solutions to these problems.

Problem 1

$B = 5 \times 10^{-2}$ weber/m²

Problem 2

$i = 18 \times 10^{-3}$ amp, counterclockwise

Problem 3

$N_S = 1900$ turns

PROGRAMMED STUDY SECTION

1 Faraday's famous experiments on induced currents consisted of one arrangement as shown to the right. Oersted had earlier shown that magnetic fields were associated with current-carrying wires. What was Faraday trying to show with this experiment?

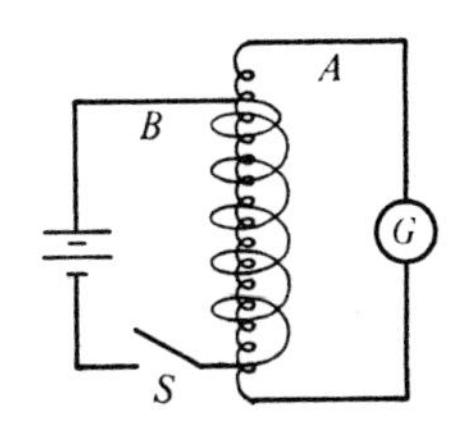

Faraday wanted to show just the converse of what Oersted had done (i.e., that currents in wires could result from the interaction of those wires with magnetic fields).

2 The galvanometer G in the circuit diagrammed in frame 1 is the device Faraday used to detect an "induced" current in the wire A. What in the circuit is the source of the magnetic field with which he sought to induce the current in A?

The coil B with the battery supplies the current which produces the magnetic field in B.

3 Faraday found that the galvanometer deflected (indicated a current in A) only during the short time that the switch S was either opened or closed. How do we interpret that fact in terms of the ability of the magnetic field in B to induce a current in A?

It is only a *changing* magnetic field that can induce a current in A. When the switch is closed the field is increasing; when the switch is opened, the magnetic field is decreasing.

4 Here we show a different experiment in which we have a constant magnetic field pointing into the paper. As you slide the wire W down, the galvanometer G deflects. What was the condition of G in Faraday's experiment when the field of B was constant?

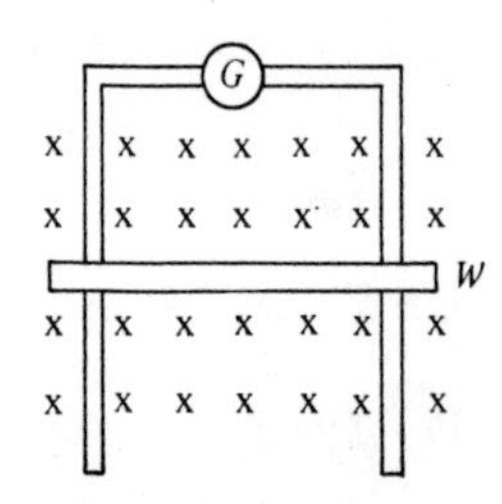

Undeflected (His experiments required a changing magnetic field because the coils were fixed as to size and position.)

5 We can explain both phenomena if we talk not about the magnetic field itself, but rather about the *flux* of the magnetic field. (You might profitably review flux in the programmed study section of the chapter on electric fields and Gauss's law.)

Similar to the definition of the flux of **E**, we can write for the flux of **B**

$$\Phi_B = \Sigma \mathbf{B} \cdot \Delta \mathbf{S}$$

We represent our two experiments below.

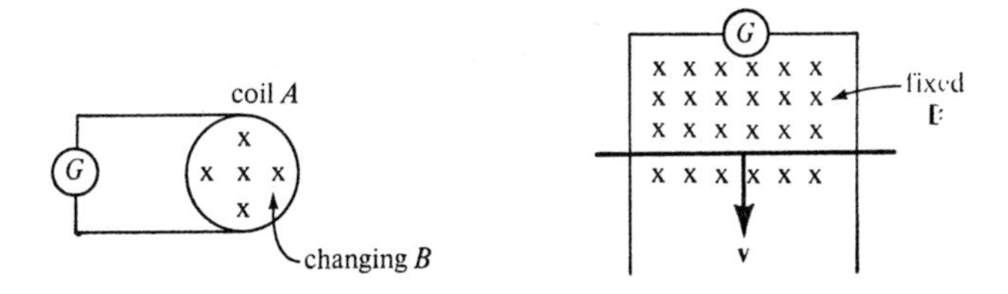

Faraday Experiment "New" Experiment

If we assert that both galvanometers deflect because of a changing flux of **B**, what in particular is causing the change in Φ_B in each case?

In Faraday's experiment **B** is changing while the area of the loop of coil A is not. (Actually **B** changes because the current producing **B** changes.) In the second experiment **B** is constant but the area is changing. Actually, the area of the loop is increasing.

6 If we speak of induced emf's instead of currents, the same idea about a changing flux of **B** still holds. Faraday's law is thus

$$\mathcal{E} = -\frac{\Delta\Phi}{\Delta t}$$

where $\Phi_B = \Sigma \mathbf{B} \cdot \Delta \mathbf{S}$

So far we have considered the following two cases:

- constant area and changing magnetic field
- constant magnetic field and changing area

From what you know about scalar products, how could we have a changing flux with a constant magnetic field and a constant area (say a loop of a certain size)?

Since the magnitude of $\mathbf{B} \cdot \Delta\mathbf{S}$ depends on the relative orientation of **B** and $\Delta\mathbf{S}$, we could obtain a change of flux by changing their relative orientation (e.g., rotating the loop). So we have three methods to change flux:

(1) Change B.
(2) Change the area ΔS.
(3) Change relative orientation of **B** and $\Delta\mathbf{S}$.

7 The previous answer is, of course, the explanation for the basic generator.

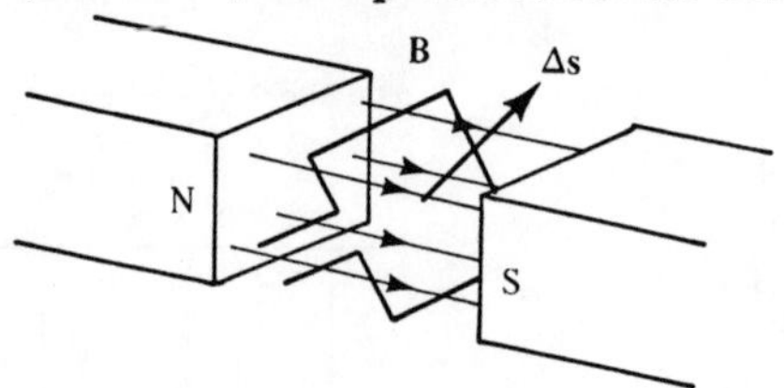

As depicted above, the area A of the loop is a constant as is the reasonably uniform **B** field produced by the two permanent magnets. However, their relative orientation is changed as we rotate the loop (armature) of this elementary generator and thus the area vector $\Delta\mathbf{S}$.

Looking at a side view and considering only a small element of area $\Delta\mathbf{S}$ we can imagine two situations as shown below.

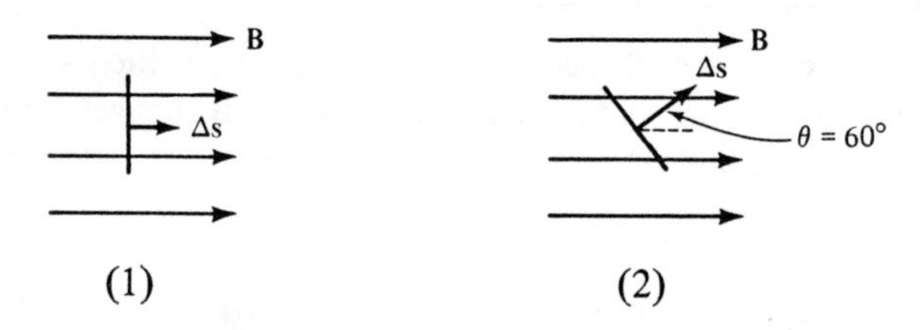

In both cases, use $B = 10^{-2}$ weber/m^2 (about what you could expect from an inexpensive bar magnet) and $A = 0.01$ m^2 to calculate the flux Φ_B.

Case 1: Φ_B = ____________ Case 2: Φ_B = ____________

Case (1):

$\Phi_B = 10^{-4}$ weber

$\Phi_B = \Sigma\mathbf{B}\cdot\Delta\mathbf{S}$
$\Phi_B = \Sigma B\Delta S \cos 0°$
$\Phi_B = B\Sigma\Delta S$, $\cos 0° = 1$
$\Phi_B = BA = 10^{-4}$ weber

Case (2):

$\Phi_B = 0.5 \times 10^{-4}$ weber

$\Phi_B = \Sigma\mathbf{B}\cdot\Delta\mathbf{S}$
$\Phi_B = \Sigma B\Delta S \cos 60°$
$\Phi_B = 0.5B\Sigma\Delta S$, $\cos 60° = \frac{1}{2}$
$\Phi_B = 0.5BA = 0.5 \times 10^{-4}$ weber

In both cases **B** is constant. In both cases $\Sigma\Delta S = A$, the area of the entire loop, and is constant.

8 What is the magnitude of the change, $\Delta\Phi_B$, in the flux of **B** through A when the loop is rotated 60° as in the previous frame?

Change in flux = final flux − initial flux
$= 0.5 \times 10^{-4}$ weber $- 1.0 \times 10^{-4}$ weber $= -0.5 \times 10^{-4}$ weber

The sign here is unimportant for present purposes. We are interested in the magnitude of the change in flux which is $\Delta\Phi_B = 0.5 \times 10^{-4}$ weber.

9 If the generator took $\frac{1}{10}$ sec to rotate through 60°, what (average) emf was induced in the loop?

$$\mathcal{E} = -\frac{\Delta\Phi_B}{\Delta t}$$

$$\mathcal{E} = \frac{\text{change in flux}}{\text{change in time}} = 5 \times 10^{-4} \text{ weber/sec}$$

Since a weber/sec is the same as a volt, $\mathcal{E} = 5 \times 10^{-4}$ volt. This is the magnitude of the average induced emf during the 0.1 second time interval.

10 Consider now the effect of an increasing or decreasing flux. The flux of **B** (increases/decreases) as the slide wire W is moved up.

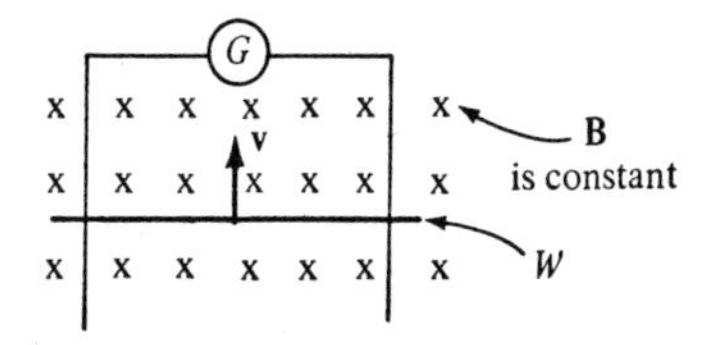

---

decreases

B remains constant while the area of the loop becomes smaller. $\Phi_B = BA$ where A is getting smaller and B is constant.

11 Lenz's law has something to say about this decreasing flux of **B**. There are many ways to state this law and some of them are confusing to the beginner. Try this one:

> Lenz's law states that "induced currents tend to oppose changes in flux."

Earlier in this section we stated three ways to change the flux of **B**. What were they?

---

(1) Change B.
(2) Change the area ΔS.
(3) Change the relative orientation of **B** and Δ**S**.

12 Which of the above three ways is being employed in frame 10 to decrease the flux of **B**?

---

(2) Change the area ΔS.

13 It seems clear that induced current in the loop can only affect one of the three options available to "oppose" this decreasing change of flux. Which is it? (Consult the answer to frame 11 again.)

---

(1) Change B. (For this somewhat rigid situation we can rule out the changing of the area or the relative orientation.)

14 According to Lenz's law in what way should the induced current change **B** in order that the decreasing flux be "opposed"?

The induced current must provide a stronger **B** through the loop to "oppose" the decrease of the flux.

15 Since the flux of **B** is decreasing, the induced current in the wire must provide an additional magnetic field **B**′ (due to the induced current) which will add to **B** and thus compensate for the decreasing area. Should the diagram be modified by indicating **B**′ as aiding or opposing **B**?

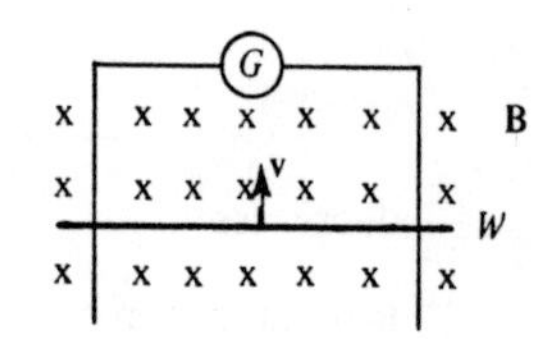

Aiding **B**

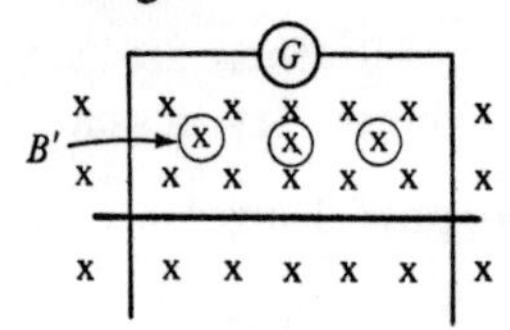

We show the x's circled to indicate that the induced current in the loop provides a magnetic field **B**′ in the same direction as **B**.

16 Lenz's law then is the following:

The magnetic flux $\Phi_B = BA$ is decreasing in this case because A is decreasing.

But the induced current in the loop due to induction, in accordance with Lenz's law, produced a magnetic field **B**′ sucn that

$$\Phi_{\text{Total}} = \underbrace{(B + B')}_{\text{increases}} \; \underbrace{A}_{\text{decreases}}$$

tending to keep the total flux constant. There is compensation occurring here. Which way must a positive induced current flow to produce the circled x's in the answer above?

Clockwise. (Curve your right-hand fingers in the direction of the current such that your thumb points in the direction of **B**′.)

SOLUTIONS TO SAMPLE PROBLEMS

Problem 1

A search coil is a small circular coil which can be used to measure magnetic fields.

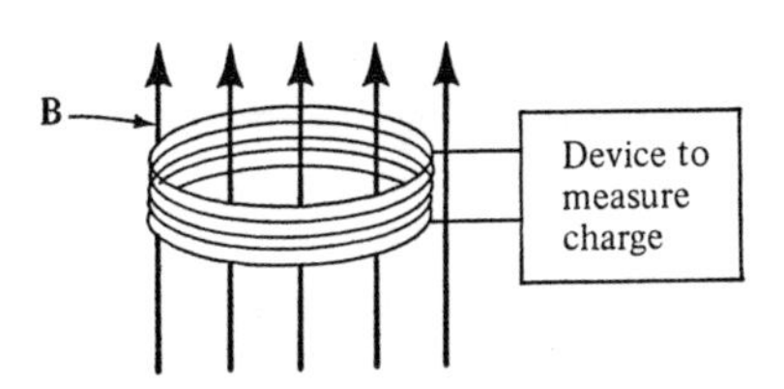

The instrument is aligned as shown with the plane of the coil perpendicular to the unknown **B** and then quickly flipped 90°. Another instrument which measures charge is attached.

Show that Q is proportional to **B** and numerically determine **B** for a search coil of the following description.

$$N = 50 \text{ turns}$$
$$A = 4 \text{ cm}^2$$
$$R = 25\ \Omega$$

where the charge measured during a flip is 4×10^{-5} coul.

1.1 A device known as a "search coil" consists (for example) of $N = 50$ turns of wire with a cross-sectional area $A = 4\text{ cm}^2$ and a resistance $R = 25\ \Omega$. Such a device can be used to determine the value of a constant magnetic field. In the sketch we show the plane of the search coil perpendicular to the unknown field. What is the flux of **B** at this time?

$\Phi_B = NBA$

For one turn, $\Phi_B = \Sigma \mathbf{B}\cdot\Delta\mathbf{S}$.
$\mathbf{B} \| \Delta\mathbf{S}$, $\cos 0° = 1$, so $\Phi_B = B\Sigma\Delta S$ since **B** is constant.
$\Sigma\Delta S = A$, $\quad \Phi_B = BA$ (for one turn)
$\Phi_B = NBA$ (for N turns)

1.2 If now we quickly flip the search coil 90°, what is the new flux of **B** through the coil?

$\Phi_B = 0$

$\Phi_B = \Sigma \mathbf{B} \cdot \Delta \mathbf{S}$
$\mathbf{B} \perp \Delta \mathbf{S}$, $\cos 90° = 0$
$\Phi_B = 0$

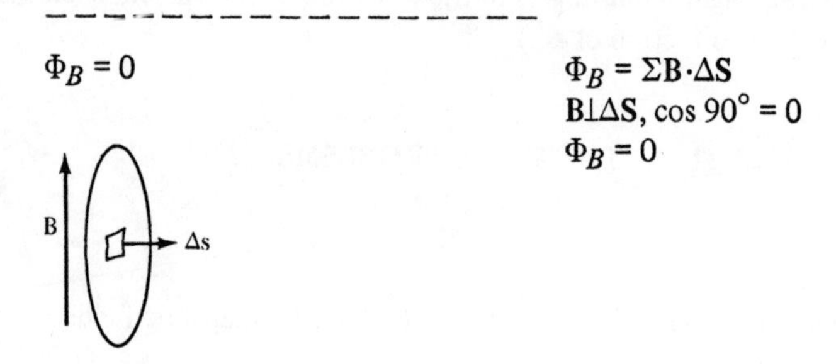

1.3 What is the change in the flux during the time Δt required to flip the coil?

$\Delta\Phi_B = NBA - 0 = NBA$ (or, if you prefer, $\Delta\Phi_B = 0 - NBA = -NBA$)

Remember that the sign has to do with the direction of the induced current. We won't worry about that aspect.

1.4 What is the induced emf in the coil?

$$\mathcal{E} = -\frac{\Delta\Phi_B}{\Delta t} = -\frac{NBA}{\Delta t}$$

1.5 Assuming the resistance to be only that of the search coil, what current is induced in the search coil during the time Δt? Express your answer in terms of B, A, R, and Δt.

$$i = \frac{\mathcal{E}}{R} = \frac{NBA}{R\Delta t}$$

1.6 Using the definition of current, rewrite the previous answer.

$$i = \frac{\Delta q}{\Delta t}$$

$$i = \frac{\Delta q}{\Delta t} = \frac{NBA}{R\Delta t}$$

$$\Delta q = \frac{NBA}{R}$$

1.7 The last answer implies that measuring Δq and knowing A, N, and R allows one to determine B. Use $R = 25\ \Omega$, $A = 4 \times 10^{-4}\ \text{m}^2$ (note the change of units) and $\Delta q = 4 \times 10^{-5}$ coul to determine B.

$B = 0.05\ \text{weber/m}^2$

$$B = \frac{(\Delta q)R}{NA}$$

$$B = \frac{4 \times 10^{-5}\ \text{coul} \times 25\Omega}{4 \times 10^{-4}\ \text{m}^2 \times 50}$$

$$B = 0.05\ \text{weber/m}^2$$

1.8 Note that it turns out that the result is independent of time. That is, the search coil may be rotated in any manner (fast, slow, not at a constant rate, etc.). All that matters is that it was flipped through the 90° angle as shown in frame 1.2.

There are several ways of measuring the charge quantitatively so that this method of measuring **B** fields is in fact used. One could, for example, measure Δq with a calibrated electroscope. Another instrument used to measure Δq is the so-called ballistic galvanometer.

1.9 One final question to focus your attention on flux changes. We calculated Δq based on a flip of 90° during which Φ_B went from BA to zero. Remembering that the induced current is determined in part by $\Delta\Phi_B = \Phi_f - \Phi_i$ (initial and final), qualitatively what would happen to Δq if the flip were

(a) 180°
(b) 360°

Think about $\Phi = \mathbf{B} \cdot \mathbf{A}$ in terms of spatial orientation before trying to answer.

(a) Twice as large; (b) Zero

To explain things we discuss answer (b) first. Since the coil ends up (360° turn) where it started, the $\Phi_f = \Phi_i$; therefore, $\Delta\Phi_B = 0$.

In part (a) $\Phi_f = -BA$ because **B** and **A** are oppositely directed. $\Phi_i = BA$ because **B** and **A** are parallel. Therefore, $\Delta\Phi_B = -BA - BA = -2BA$.

Problem 2

The wire W of length $\ell = 0.2$ m moves to the right with constant velocity $v = 0.3$ m/sec. If the resistance of the loop is approximately fixed at 10 Ω and $B = 3.0\ \text{weber/m}^2$, what is the magnitude and direction of the induced current?

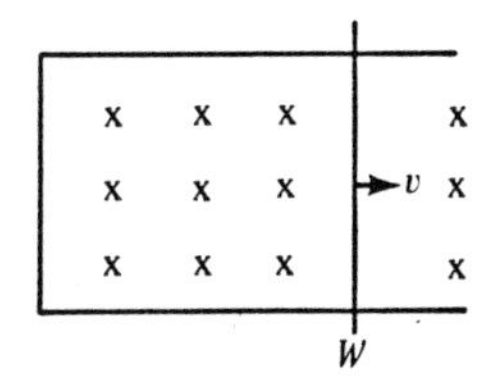

2.1 We will approach this problem from the point of view of changing magnetic flux, but the result will also permit discussion of emf's induced in moving wires which are not part of a closed loop. In the diagram, the wire W is moving with constant velocity v as shown. The remaining wire is fixed. Good electrical contact is, of course, maintained between both wires.

The magnetic field **B** is constant and directed into the paper. What is the reason for the flux of **B** to change?

$$\Phi_B = \Sigma \mathbf{B} \cdot \Delta \mathbf{S}$$

The area of the closed loop (rectangle actually) is changing.

2.2 Is the area increasing or decreasing?

Increasing

2.3 Is the flux of **B** increasing or decreasing?

Increasing

2.4 To determine the induced emf, Faraday's law of induction

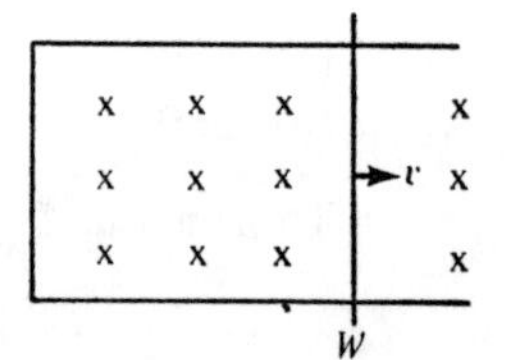

$$\mathcal{E} = -\frac{\Delta \Phi}{\Delta t}$$

requires that we determine how this increasing flux of **B** changes with time. Since we know the flux change is due to an area change we can concentrate on how the area changes. Obviously we are interested in how much the area changes in a time Δt. Modify the sketch to show the position of W after a time Δt.

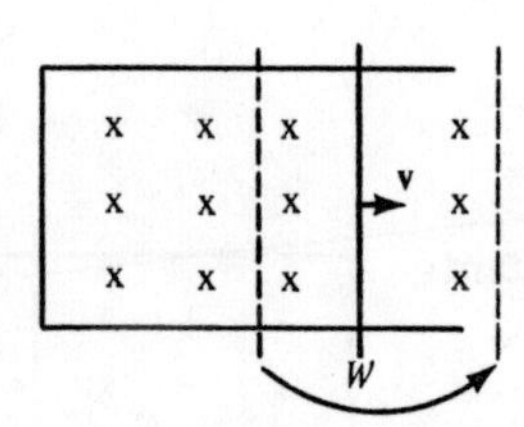

The wire W has moved to the right (in the direction of v) as shown.

2.5 By how much did the area change as shown in the previous answer? Draw or describe the area in terms of the picture.

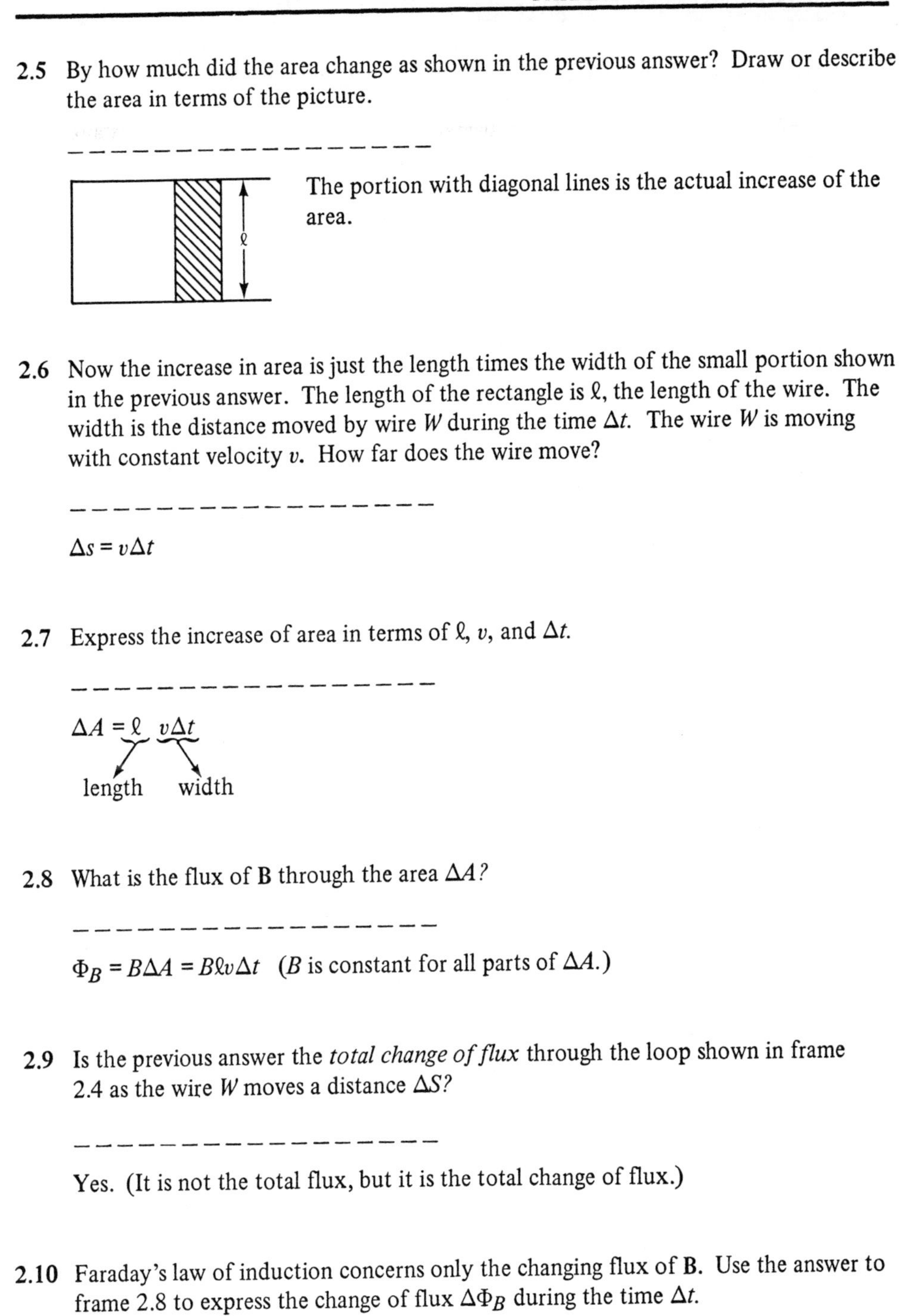

The portion with diagonal lines is the actual increase of the area.

2.6 Now the increase in area is just the length times the width of the small portion shown in the previous answer. The length of the rectangle is ℓ, the length of the wire. The width is the distance moved by wire W during the time Δt. The wire W is moving with constant velocity v. How far does the wire move?

$\Delta s = v\Delta t$

2.7 Express the increase of area in terms of ℓ, v, and Δt.

$\Delta A = \ell \; v\Delta t$

length (↙ ℓ)   width (↘ $v\Delta t$)

2.8 What is the flux of **B** through the area ΔA?

$\Phi_B = B\Delta A = B\ell v\Delta t$ (B is constant for all parts of ΔA.)

2.9 Is the previous answer the *total change of flux* through the loop shown in frame 2.4 as the wire W moves a distance ΔS?

Yes. (It is not the total flux, but it is the total change of flux.)

2.10 Faraday's law of induction concerns only the changing flux of **B**. Use the answer to frame 2.8 to express the change of flux $\Delta\Phi_B$ during the time Δt.

$\Delta\Phi_B = B\ell v\Delta t$

2.11 Write an expression for the emf $\mathcal{E}$ induced during the time Δt.

$$\mathcal{E} = -\frac{\Delta\Phi_B}{\Delta t} = \frac{B\ell v\Delta t}{\Delta t} = B\ell v$$

2.12 The data from the problem is

$B = 3$ weber/m^2 $\quad \ell = 0.2$ m
$v = 0.3$ m/sec $\quad R = 10\ \Omega$

Obtain numerically the current in the loop.

$i = 18 \times 10^{-3}$ amp

$$i = \frac{\mathcal{E}}{R} = \frac{B\ell v}{R}$$

$$i = \frac{3 \times 0.3 \times 0.2}{10} = 18 \times 10^{-3} \text{ amp}$$

2.13 Which direction must the current move in order to satisfy Lenz's law? Look at the answer to frame 2.4 and think in terms of counter-acting the flux change.

Counterclockwise (The flux is increasing so the induced current adds a magnetic field which must oppose the field into the paper.)

2.14 The remaining frames will discuss another interpretation of $\mathcal{E} = B\ell v$. We can think in terms of the magnetic field in

$$F = qvB \sin\theta$$

as acting on the mobile charges in the conduction wire of length ℓ. The force acting on the charge $+q$ depends upon its velocity (which is the same as the wire) v, B, and the angle θ between **v** and **B**. What is the sin θ for the situation sketched?

ℓ
v
B into paper

$\sin\theta = 1$ (**v** and **B** are perpendicular.)

2.15 Considering the mobile charges to be positive, $F = qvB$. What is the direction of **F** as given by the right-hand rule?

Toward the top of the wire.

2.16 This force **F** separates the charges in the wire as shown. This constitutes an emf.

$$\mathcal{E} = B\ell v$$

The force **F** separates charge in the moving wire until equilibrium is reached between the magnetic force due to motion and the electrical forces due to charge concentration.

Problem 3

A coil in the form of a long solenoid has 1000 turns with a cross-sectional area of 0.1 m^2. A separate coil of similar cross-section is wound over the primary coil. An emf of 6 volts is induced in the secondary coil if the current in the primary changes by 0.5 ampere in 0.02 sec. How many turns are used for the secondary?

3.1 We show the primary and secondary coils of different size for clarity. Which coil is the source of a changing magnetic field which gives rise to $\Delta\Phi_B$ for the secondary?

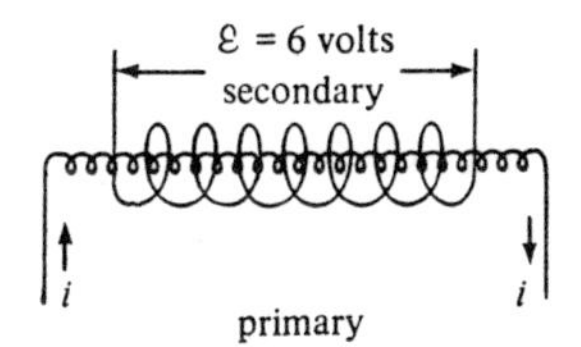

The primary coil

3.2 With the current in the primary coil as shown, indicate the direction of the magnetic field of the primary solenoid. The current is conventionally positive.

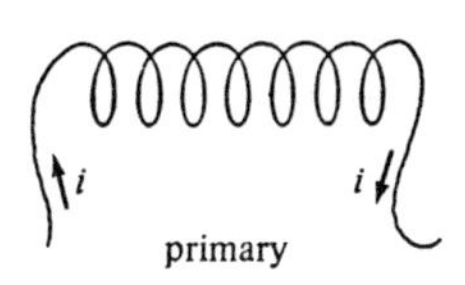

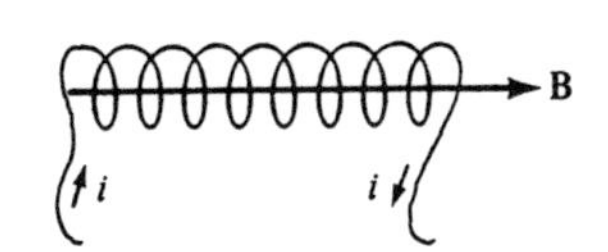

3.3 This **B**, the magnetic field produced by the primary solenoid, is given by the expression

$$B = \mu_0 N_p i, \qquad \text{where } N_p = \text{turns of primary coil}$$
$$i = \text{current in primary coil}$$
$$\mu_0 = 4\pi \times 10^{-7} \text{ web/amp-m}$$

At the instant the current in the primary is i, what is the flux of **B** through the secondary solenoid? For the moment imagine the secondary to be one loop.

$\Phi_B = BA = \mu_0 N_p i A$

For the concentric solenoids the magnetic field of the primary also has the same direction through the secondary. B is constant, therefore

$$\Phi_B = \underbrace{\mu_0 N_p i}_{B \text{ of primary}} \quad \underbrace{A}_{\text{area of secondary coil}}$$

3.4 What is the flux through the secondary coil for N_s secondary turns?

$$\Phi_B = \underbrace{\mu_0 N_p i}_{B \text{ of primary}} \quad \underbrace{N_s}_{N \text{ of secondary}} \quad \underbrace{A}_{\text{area of secondary}}$$

3.5 Which term of the previous answer could be conveniently changed in order to change the flux Φ_B in the secondary?

----

i (the current of the primary)

3.6 From the problem i changes 0.5 amp in 0.02 sec. Write an expression for

$$\mathcal{E} = -\frac{\Delta\Phi_B}{\Delta t} = \underline{\qquad\qquad\qquad\qquad}$$

for the secondary. Use the answer to frame 3.4 and the information of this frame.

$$\mathcal{E} = -\mu_0 N_p N_s A \,\frac{0.5 \text{ amp}}{0.02 \text{ sec}} \qquad \mathcal{E} = -\frac{\Delta\Phi_B}{\Delta t} = -\frac{\mu_0 N_p N_s A\,\Delta i}{\Delta t}$$

3.7 Solving the previous answer for the unknown and ignoring the sign we have

$$\frac{\mathcal{E}\,(0.02 \text{ sec})}{\mu_0 N_p A\;(0.5 \text{ amp})} = N_s$$

Obtain a numerical result using data from the problem.

$N_s = 1900$ turns

Using only the numbers:

$$N_s = \frac{6 \times 0.02}{4\pi \times 10^{-7} \times 1000 \times 0.1 \times 0.5}$$

And in powers of ten:

$$N_s = \frac{6 \times 2 \times 10^{-2}}{4\pi \times 10^{-7} \times 10^{3} \times 5 \times 10^{-2}}$$

$N_s = 1900$ turns

SELF–TEST

1 A rectangular coil 10 cm by 20 cm is made of 100 tightly wound turns. The coil is arranged with its plane initially parallel to the magnetic field between the magnets. A uniform magnetic field of 0.4 weber/m^2 exists between the magnets. If the coil is rotated 90° in 0.2 sec, what is the magnitude of the average emf induced in the coil?

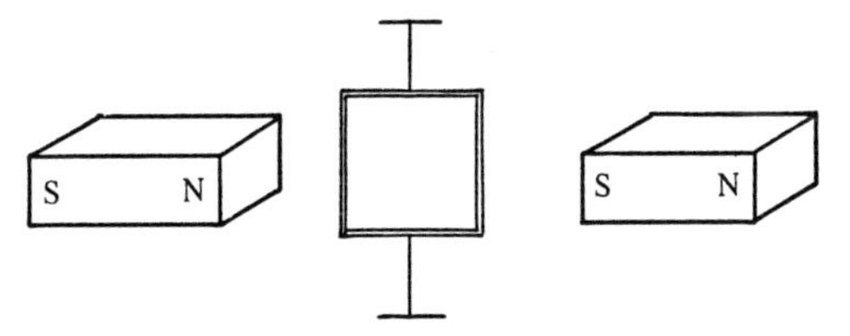

2 Looking down on the situation described in problem 1, consider the initial condition depicted here. The plane of the coil as shown makes an angle of 30° with direction of the magnetic field. If the coil is rotated CCW 60° in 0.2 sec, what is the magnitude of the average induced emf?

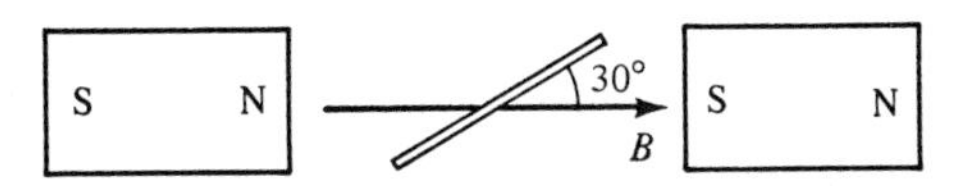

3 A commercial jet airplane has a wing-span of approximately 200 m. Flying coast to coast the wing is a conductor moving through a vertical magnetic field 5×10^{-5} weber/m^2. What emf is induced tip to tip if the aircraft flies at 600 mph?

4 A magnet is thrust towards the center of a coil of 4 turns such that the magnetic flux through the coil changes from 0.32 weber to 0.51 weber in 0.01 sec. If the coil has a resistance of 20 Ω, what is the magnitude and direction (as viewed from the side near the magnet) of the induced current?

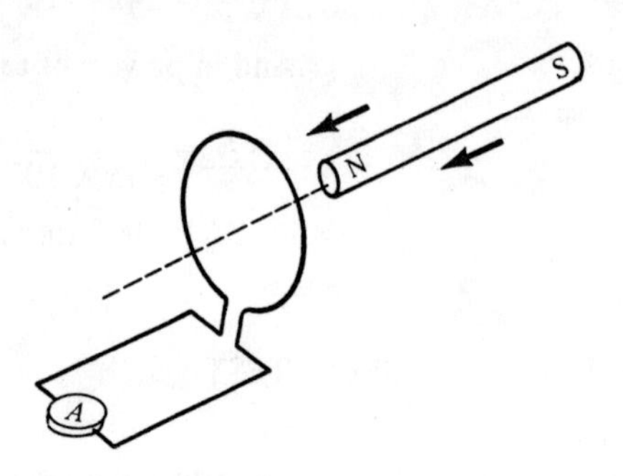

Answers to Self-Test

1 400 volts

2 200 volts

3 2.7 volts

4 3.8 amp, CCW

CHAPTER SIX
Current Electricity

If the sample problems and objectives below identify your weak points, go directly to the programmed study section on page 99. If not, try the problems and compare your answers with those that follow. If you can do all the problems easily and if you are familiar with the objectives, you may wish to skip all or part of this chapter. The programmed study section covers techniques and concepts basic to solving the sample problems and fulfilling the objectives in this chapter. A programmed, step-by-step solution of each sample problem begins on page 104. A self-test is included at the end of the chapter.

SAMPLE PROBLEMS AND OBJECTIVES

Problem 1

Show that the equivalent resistance R_T of two resistors R_1 and R_2 in parallel is given by the expression

$$R_T = \frac{\text{product of resistors}}{\text{sum of the resistors}}$$

Use the above idea to find the equivalent resistance of the network shown to the right.

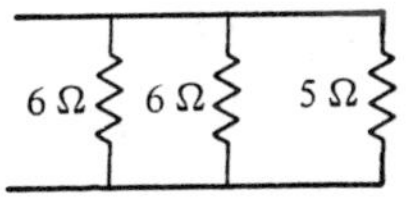

Objectives:

1. Reviewing the algebra of adding fractions.
2. Describing techniques of circuit decomposition useful in solving problems.

Problem 2

Calculate the current through each resistor and the potential difference across each resistor.

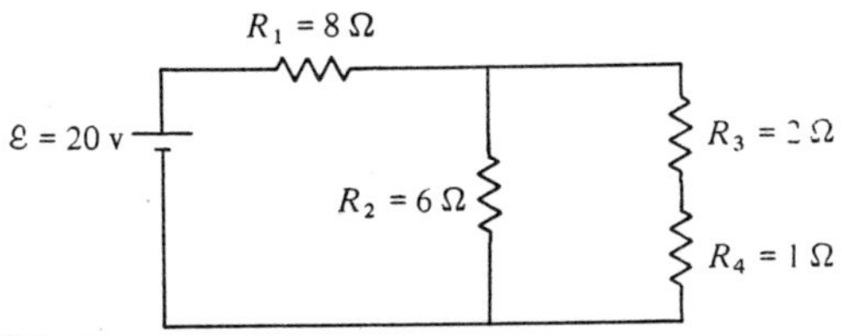

Objectives:

1. Practice at decomposing resistance circuits into an equivalent resistance.
2. Applying Kirchhoff's rules to various parts of a circuit.

Problem 3

Determine the current through each resistor.

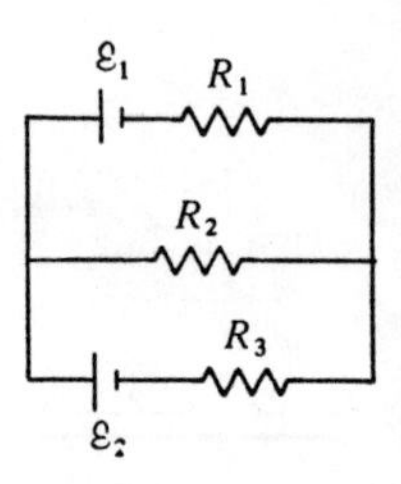

$\mathcal{E}_1 = 5$ volt
$\mathcal{E}_2 = 10$ volt
$R_1 = 2\ \Omega$
$R_2 = 4\ \Omega$
$R_3 = 3\ \Omega$

Objectives:

1. Discussing Kirchhoff's circuit rules as applied to multiloop circuits.
2. Solving problems involving simultaneous equations.

Answers to Sample Problems

See page 104 for programmed, step-by-step solutions to these problems.

Problem 1

$1.8\ \Omega$

Problem 2

R_1 - 2 amp, 16 volts

R_2 - $\frac{2}{3}$ amp, 4 volts

R_3 - $1\frac{1}{3}$ amp, $\frac{8}{3}$ volts

R_4 - $1\frac{1}{3}$ amp, $\frac{4}{3}$ volts

Problem 3

current in R_1 = 0.1 amp
current in R_2 = 1.3 amp
current in R_3 = 1.4 amp

PROGRAMMED STUDY SECTION

This chapter is rather practical in nature. The main thrust is to provide an organized approach to the determination of variables in direct current circuits. The programmed study section covers the high points of the rules of circuit behavior.

1 The first few frames discuss circuits similar to old and new Christmas tree lights. A careful discussion of these two familiar situations should settle many of the confusing points in circuit problems.

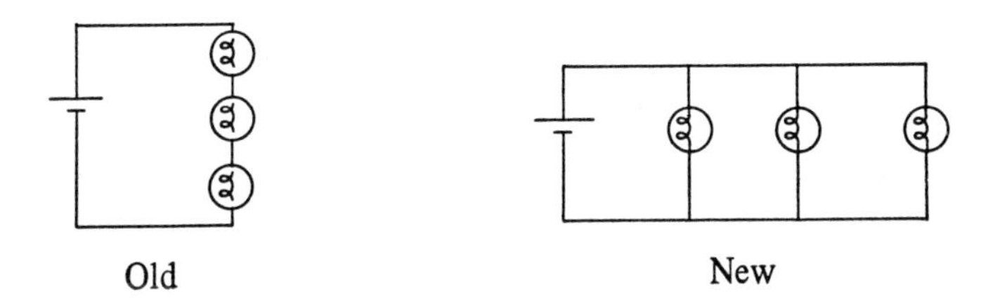

In which of the circuits above can one remove a bulb without affecting the remaining bulbs?

The parallel connection (which is labeled "new")

2 In the series case (labeled "old") removing one bulb would break the circuit, thus none would light. All electrical objects have at least two terminals (e.g., batteries have two ends, toaster plugs have at least two prongs, filaments which glow in light bulbs have two ends, lamp cords have two wires, etc.). To connect bulbs to a source of electrical power requires that both ends of the filament be connected (directly or indirectly) to the source.

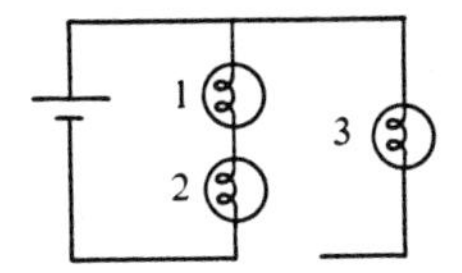

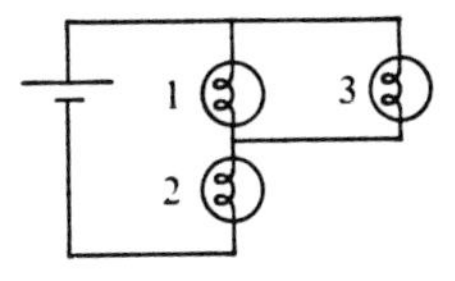

Will bulb 3 above light in each case?

No. Only one end of the filament is connected in the curcuit to the left.

3 With the idea of two terminals we can see why bulbs in parallel behave independently.

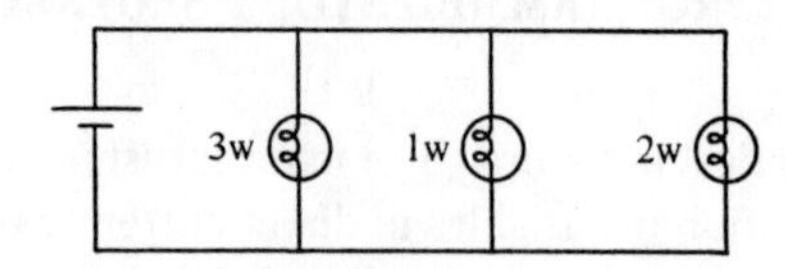

What happens to the 3 w and 2 w bulbs if the 1 w bulb burns out? (The abbreviation w stands for watt.)

Nothing. Both ends of the 3 w and 2 w bulbs are still connected to the battery, thus they are undisturbed.

4 Your home must be connected as a large parallel circuit since the kitchen light doesn't change when the bathroom light burns out or is turned off. If the bathroom light is on, does turning on the kitchen light normally make the bathroom light dimmer (or brighter) than usual?

No. If it does you should call an electrician.

5 It appears that the amount of electrical power in a parallel circuit doesn't depend on the battery alone.

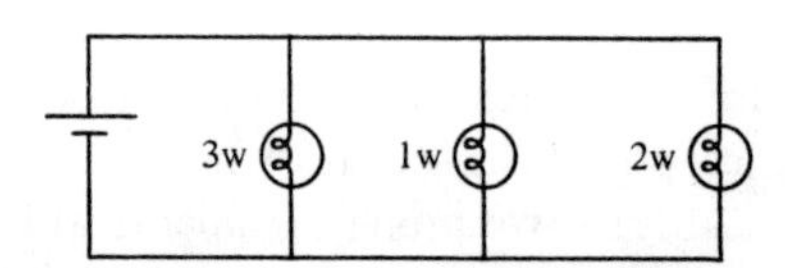

To a good approximation the terminal potential difference (voltage) of the battery remains roughly fixed as you add or remove bulbs. The total power available can be 1, 2, 3, 4, 5, or 6 watts, depending on whether all or some of the bulbs are connected. Thus, the total power may vary. What *is* different in the circuit for the different total power situations?

The total current supplied by the battery changes in each case.

6 The previous answer is represented diagrammatically below.

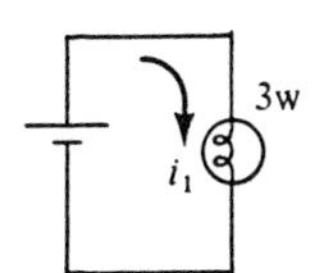

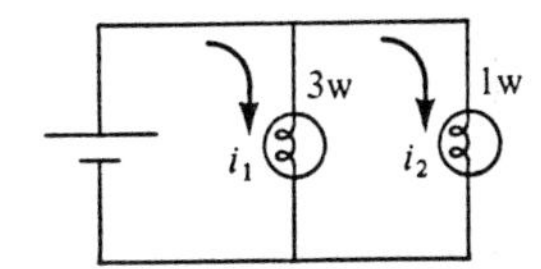

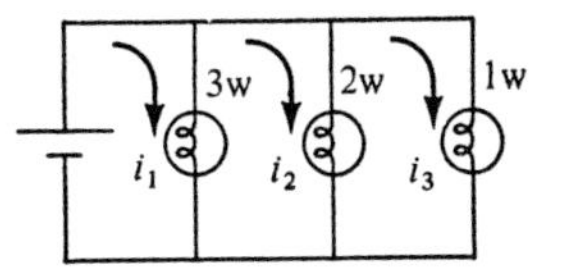

(a) The battery provides a current i_1.
(b) The battery provides a current i_1 plus another current i_2.
(c) The battery provides i_1, i_2, and i_3.

In case (c), is it necessary that i_1, i_2, and i_3 be the same?

No. The only requirement is that the battery cause a total current $i_1 + i_2 + i_3$. This differs from a series circuit where currents are all the same.

7 Kirchhoff's rule for currents requires that the algebraic sum of the currents at the junction P be zero ($\Sigma i = 0$). A junction is any circuit point where current(s) enter and leave. Using the convention that currents into a junction are positive, write an equation satisfying Kirchhoff's current rule at the junction market P. i_B is the current in the wire connected between the battery and P.

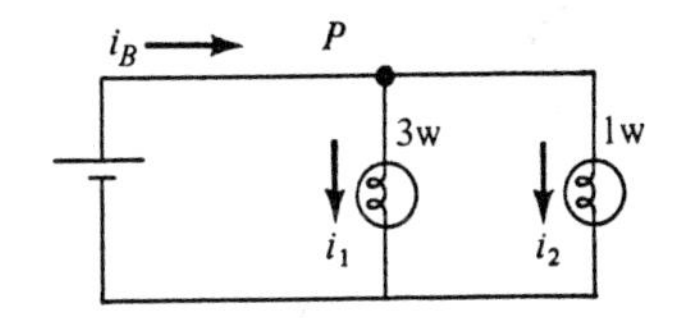

$i_B - i_1 - i_2 = 0$

8 Invoke the same rule on the point P of the circuit to the left. Comment on the resulting relationship between i_1 and i_2, the currents through each lamp.

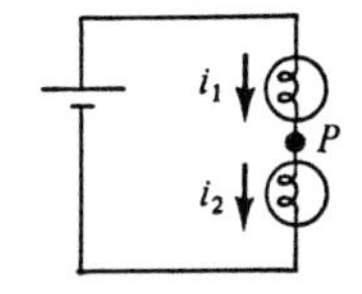

$\Sigma i = i_1 - i_2 = 0$ (or $i_1 = i_2$)

This result would obtain for any point in this circuit. The implication for this result is that for a series circuit the current entering a point is the same as the current leaving that point. In other words, the current is everywhere the same in a series circuit.

9 Obtain an expression for i_x.

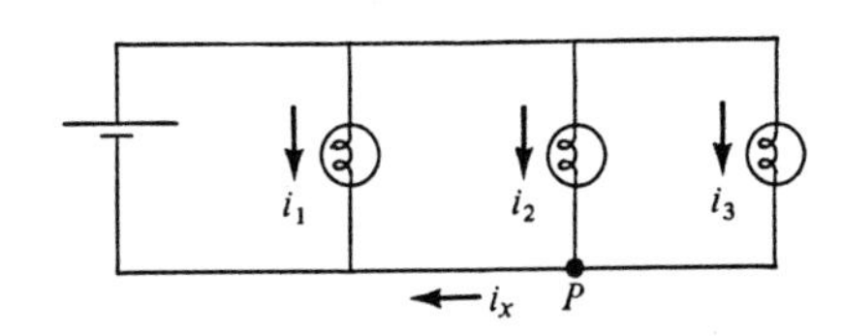

$i_1 + i_3 - i_x = 0$ (at P, $i_x = i_2 + i_3$)

10 Use Kirchhoff's current rule to determine the reading of meters B, D, and E given that A reads 10 amp, C reads 2 amp, and F reads 5 amp.

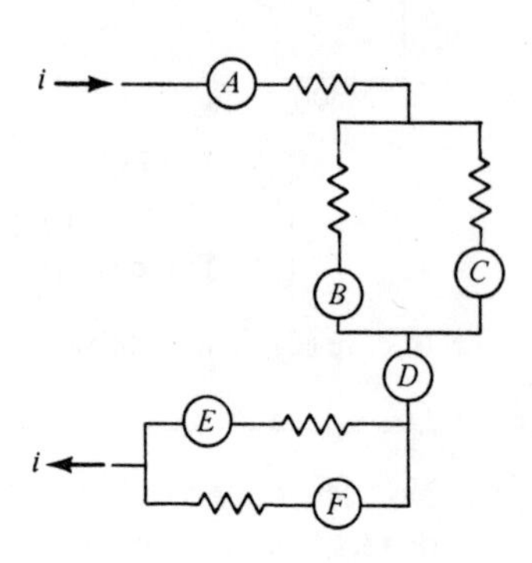

B reads 8 amp; D reads 10 amp;
E reads 5 amp

Example:

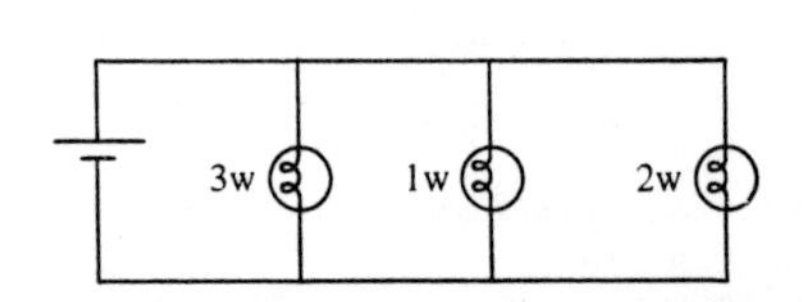

11 As we noted earlier, the battery in this circuit could deliver more or less power because the total current in the complete circuit could vary even though the circuit voltage didn't. What characteristic of each bulb determines the current passing through it?

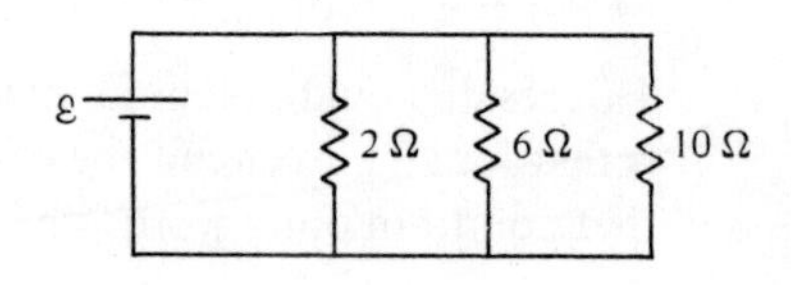

The filament resistance

12 Which of these resistive elements would have the greatest current?

$\mathcal{E}$ — 2 Ω — 6 Ω — 10 Ω

---

2 Ω (The current is inversely proportional to the resistance.)

13 In the circuit of frame 12 all resistors are connected directly to the battery so each has a potential difference equal to the electromotive force $\mathcal{E}$ of the battery. We are ignoring any internal resistance of the battery.

Ohm's law gives the current through each as $i = \mathcal{E}/R$ where $\mathcal{E}$ is *the same for each circuit element in parallel,* but R in this case is different. For $\mathcal{E} = 30$ volt, what is the current through each and the total current in the circuit?

---

15 amp through the 2 Ω circuit element
5 amp through the 6 Ω circuit element
3 amp through the 10 Ω circuit element
23 amp total

14 If we use Kirchhoff's rule for currents we have

$$i_{\text{Total}} = i_{2\,\Omega} + i_{6\,\Omega} + i_{10\,\Omega}$$

Using Ohm's law we have

$$\frac{\mathcal{E}_{\text{Total}}}{R_{\text{Total}}} = \frac{\mathcal{E}_{\text{Total}}}{R_{2\,\Omega}} + \frac{\mathcal{E}_{\text{Total}}}{R_{6\,\Omega}} + \frac{\mathcal{E}_{\text{Total}}}{R_{10\,\Omega}}$$

$$23 \text{ amp} = 15 \text{ amp} + 5 \text{ amp} + 3 \text{ amp}$$

This reduces to a familiar result for parallel circuits.

$$\frac{1}{R_T} = \frac{1}{R_{2\,\Omega}} + \frac{1}{R_{6\,\Omega}} + \frac{1}{R_{10\,\Omega}}$$

15 The description of the series circuit in frame 8 showed that the current was the same everywhere. Kirchhoff's voltage rule states that the algebraic sum of the changes in potential around a closed path is zero. In simple circuits we consider emf's as leading to an increase in potential and resistive elements as giving rise to potential drop (called iR drops).

Proceeding around the closed path from A to B to C to D and back to A we have $-iR_1 - iR_2 - iR_3 + \mathcal{E} = 0$. This is usually written

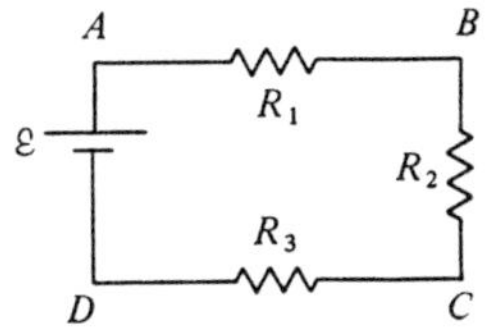

A B R_1 $\mathcal{E}$ R_2 R_3 D C

$$\mathcal{E} = iR_1 + iR_2 + iR_3$$

Is i the same for each resistive element?

Yes (This is a series circuit)

16 If we think of the resistive elements as being in a black box then we can say it has some effective total resistance R_T.

$$\frac{\mathcal{E}}{i} = R_1 + R_2 + R_3$$

Writing Kirchhoff's voltage rule as above the lefthand term has the units of resistance. $\mathcal{E}$ is the total voltage applied to the circuit and i is the total current in the circuit.

$\mathcal{E}/i$ is called ______________________

R_T, the total resistance in the circuit

(Thus, $R_T = R_1 + R_2 + R_3$, a familiar result for resistors in series.)

SOLUTIONS TO SAMPLE PROBLEMS

Problem 1

Show that the equivalent resistance R_T of two resistors R_1 and R_2 in parallel is given by the expression

$$R_T = \frac{\text{product of resistors}}{\text{sum of the resistors}}$$

Use the above idea to find the equivalent resistance of the network shown to the right.

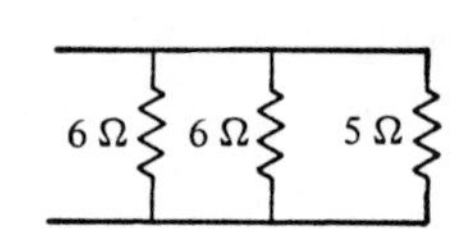

1.1 The purpose of this problem is simply to show you some useful tricks that make problem solving a little more manageable. You have seen that

$$\frac{1}{R_T} = \frac{1}{R_1} + \frac{1}{R_2}$$

is the relationship giving total resistance for two resistors in parallel. Numerically a similar example is

$$\frac{1}{2} = \frac{1}{3} + \frac{1}{6}$$

What is the common denominator of the numerical example?

---

6 (A quick way to find the common denominator for several fractions is to multiply the denominators together. For example 36 would be as appropriate as 6 in the example above.)

1.2 What is the common denominator of the first equation of frame 1.1?

$R_T R_1 R_2$ (the product of all denominators)

1.3 Multiply the equation

$$\frac{1}{R_T} = \frac{1}{R_1} + \frac{1}{R_2}$$

by its common denominator and solve that equation for R_T.

$$R_1 R_2 = R_2 R_T + R_1 R_T$$
$$R_1 R_2 = R_T(R_1 + R_2)$$
$$R_T = \frac{R_1 R_2}{R_1 + R_2} = \frac{\text{product of two resistors}}{\text{sum of two resistors}}$$

This is an easily remembered way to find the equivalent resistance of two resistors in parallel.

1.4 Use the rule of the previous frame to combine the two parallel 6 Ω resistors.

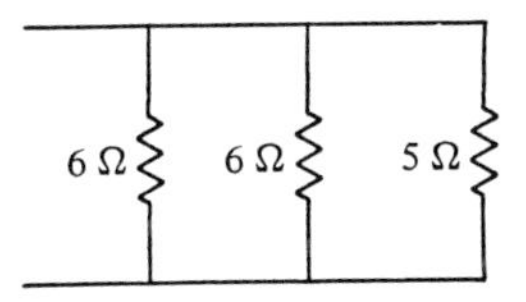

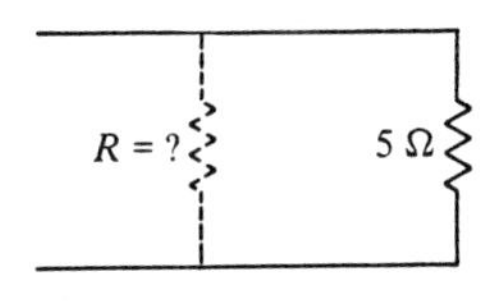

---

3 Ω 5 Ω

$$R_T = \frac{6\ \Omega \times 6\ \Omega}{12\ \Omega} = 3\ \Omega$$

Incidently if $R_1 = R_2$ then the product-over-sum rule takes an especially useful form.

$$R_T = \frac{R \times R}{R + R} = \frac{R^2}{2R} = \frac{R}{2}$$

Here the result is just the resistance of one divided by the number N of resistors. This is true no matter how many resistors there are, provided they are all the same. For example:

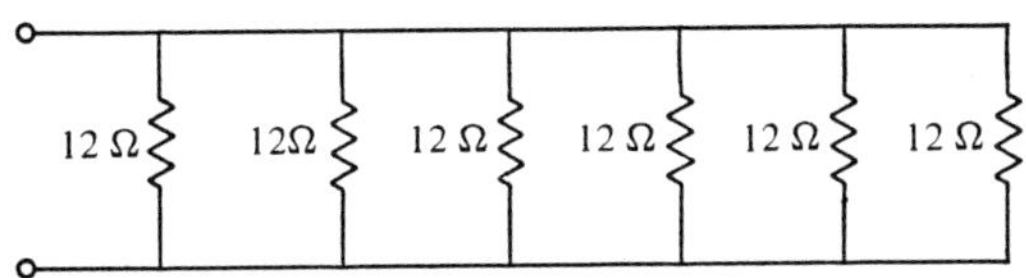

All resistors (total of 6) are the same; each has a resistance of 12 Ω.

$$R_T = \frac{R}{N} = \frac{12\ \Omega}{6} = 2\ \Omega$$

1.5 We can use the rule again having replaced the two 6 Ω resistors by their 3 Ω equivalent. Combine again.

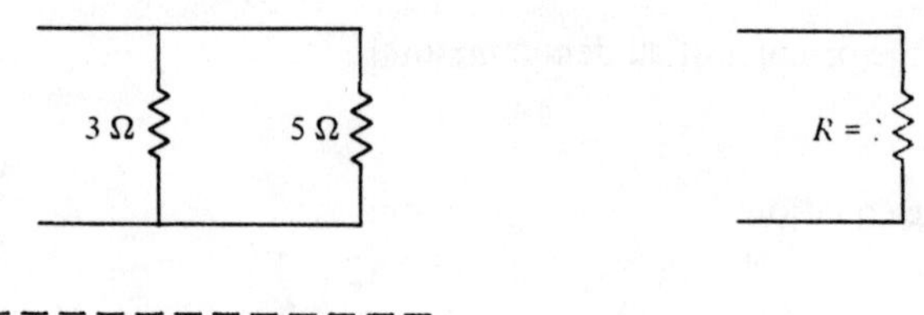

$$R_T = \frac{3\ \Omega \times 5\ \Omega}{8\ \Omega} = 1.8\ \Omega$$

1.6 Find the equivalent resistance below. (You should be able to do it in no more than two steps.)

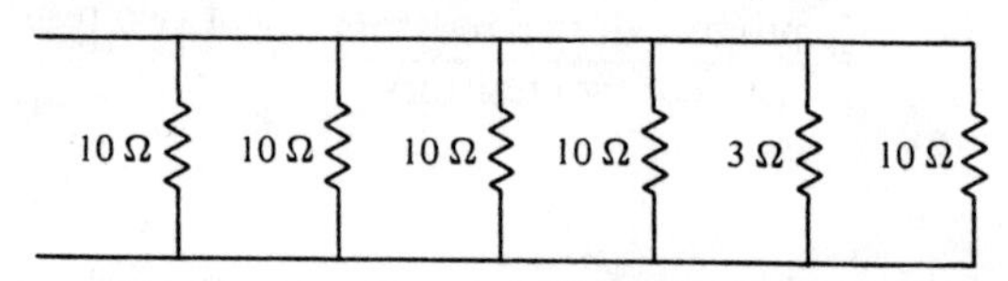

R_T = ____________ ohm

$R_T = 1.2\ \Omega$

The five 10 Ω resistors in parallel have a total resistance of 2 Ω.

$$R_T = \frac{10\ \Omega}{5} = 2\ \Omega$$

Now the circuit reduces to

2 Ω 3 Ω $= \dfrac{2\ \Omega \times 3\ \Omega}{5\ \Omega} = 1.2\ \Omega$

Problem 2

Calculate the current through each resistor and the potential difference across each resistor.

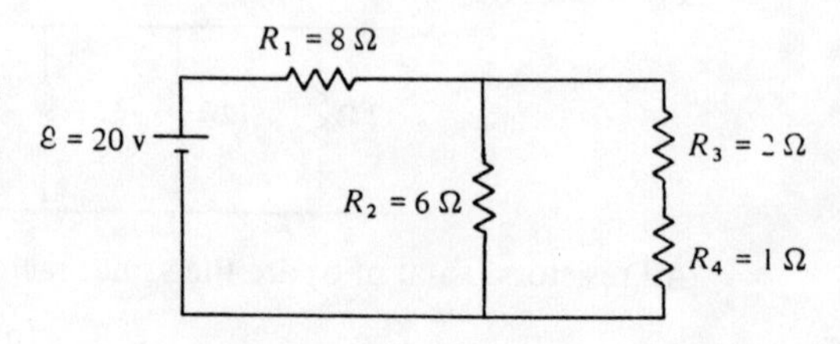

2.1 This problem is an example of simple series-parallel combinations. A good method for solving such a problem is to first reduce it to a single equivalent resistor and then build it back up, keeping track of currents according to Kirchhoff's rule. At first glance which resistors are most easily combined?

$R_3 = 2\ \Omega$

$= 3\ \Omega$

$R_4 = 1\ \Omega$

This is a series combination. Try to remember to look for things that are *obviously* in either series or parallel.

2.2 Electrically the circuit shown here is exactly the same as that of the original problem. There is now evident another simple combination. Reduce the circuit further.

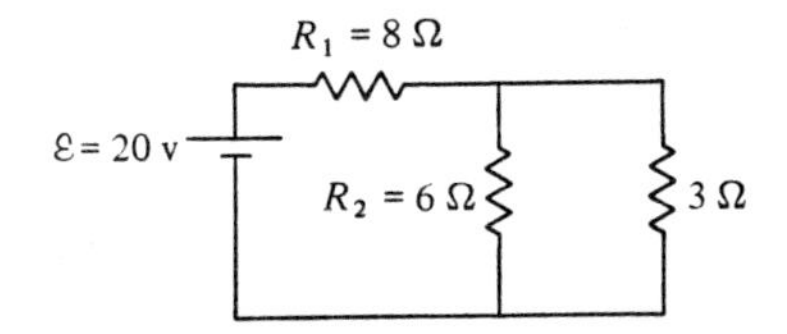

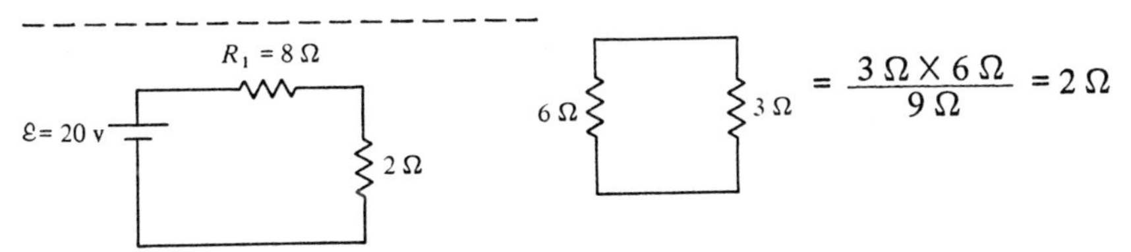

$$= \frac{3\ \Omega \times 6\ \Omega}{9\ \Omega} = 2\ \Omega$$

This is the product-over-sum technique.

2.3 Make the final simplification of the circuit.

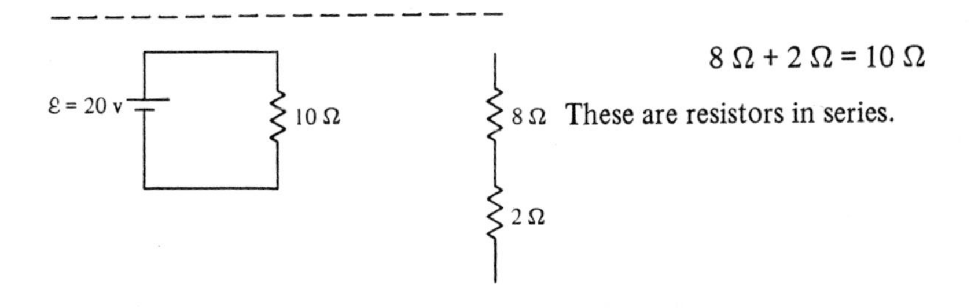

$$8\ \Omega + 2\ \Omega = 10\ \Omega$$

These are resistors in series.

2.4

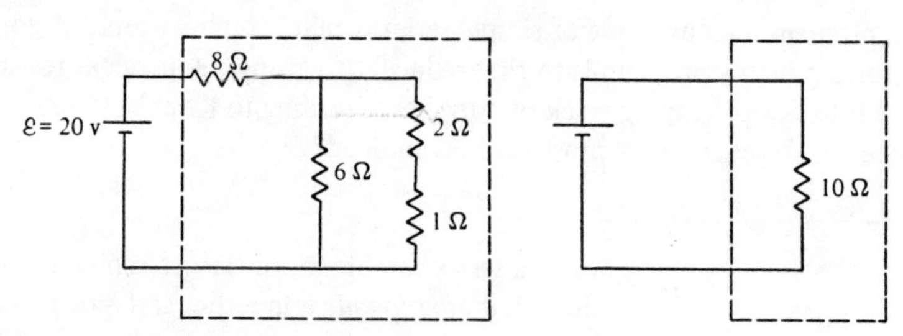

To say that these circuits are equivalent is to say that electrically the stuff inside each dotted box behaves in the same manner.

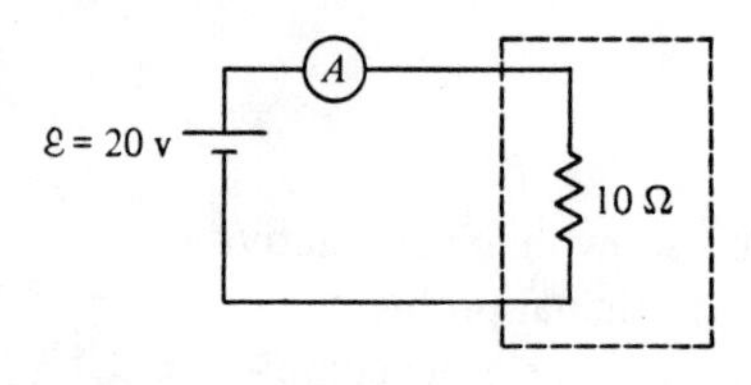

Using Ohm's law we can determine the current which would register on the ammeter.

$i =$ ________________ amp

$$i = \frac{\mathcal{E}}{R} = \frac{20 \text{ v}}{10 \ \Omega} = 2 \text{ amp}$$

2.5 The fact of electrical equivalence requires that the meter shown here read the same as the answer to the previous frame. In what circuit configuration is the ammeter (A) connected to R_1?

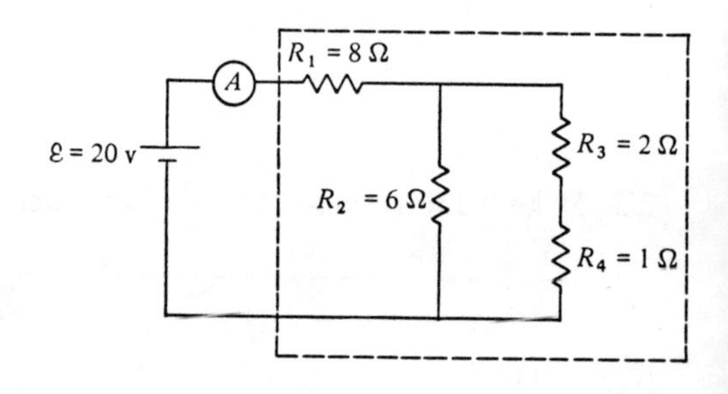

Series

2.6 If (A) reads 2 amp, what is the current through R_1?

Also 2 amp, (A) and R_1 are connected in series.

2.7 We show with arrows the currents into and out of point *P*.

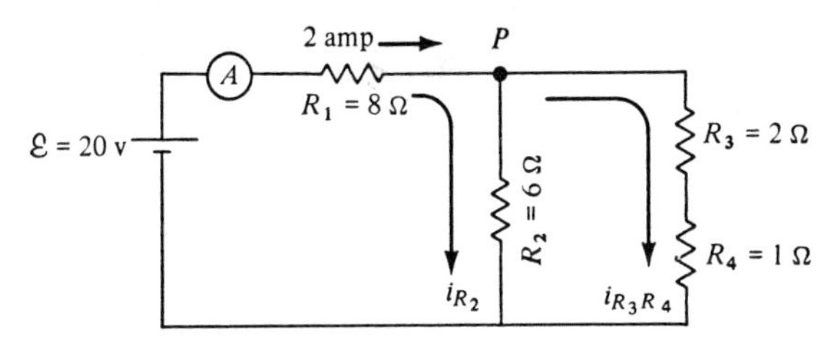

Kirchhoff's current rule requires

$$2 \text{ amp} = i_{R_2} + i_{R_3R_4}$$

This is one equation with two unknowns so we need another relationship. What is the circuit configuration of R_2, R_3, and R_4?

R_2 in parallel with the R_3-R_4 series combination

2.8 In the programmed study section we discussed the fact that electrical objects connected in parallel have the same potential difference across each parallel branch. If (V_1) voltmeter reads 4 volts, what will (V_2) read?

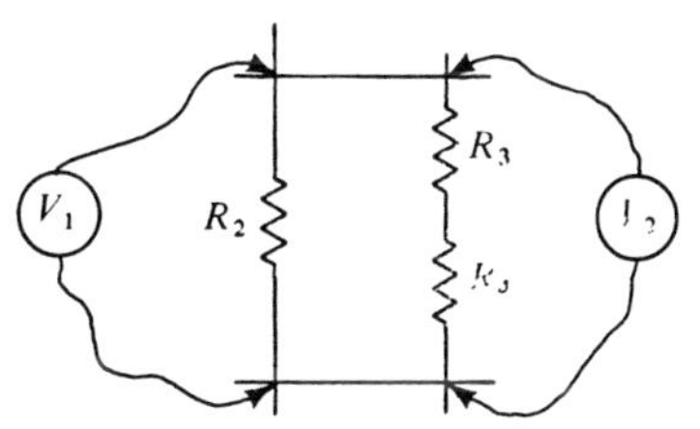

Also 4 volts (R_2 is in parallel with the R_3-R_4 series combination.)

2.9 Ohm's law states that the potential difference across resistive elements is $V = iR$ where i is the current and R the resistance. Write an equation which reflects the answer of the previous frame. Use the sketch shown here to designate variables.

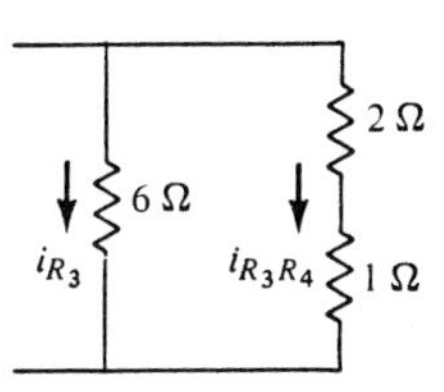

$$6i_{R_2} = 2i_{R_3R_4} + 1i_{R_3R_4} = 3i_{R_3R_4}$$

2.10 Using the results of frame 2.7 and 2.9 we have two equations and two unknowns.

$$2 = i_{R_2} + i_{R_3R_4}$$
$$6i_{R_2} = 3i_{R_3R_4}$$

(a) i_{R_2} = __________ amp

(b) $i_{R_3R_4}$ = __________ amp

(a) $\frac{2}{3}$ amp; (b) $1\frac{1}{3}$ amp

Note that the two answers add up to 2 amp as required by the first equation. Also

$$6 \times \frac{2}{3} = 3 \times 1\frac{1}{3}$$

as required by the second equation.

2.11 We now know the current through all resistors and can use Ohm's law to calculate the potential difference across each.

$R_1 = 8\ \Omega$ has a current of 2 amp.
$R_2 = 6\ \Omega$ has a current of $\frac{2}{3}$ amp.
$R_3 = 2\ \Omega$ has a current of $1\frac{1}{3}$ amp.
$R_4 = 1\ \Omega$ has a current of $1\frac{1}{3}$ amp.

Thus,

$$V_{R_1} = 2 \text{ amp} \times 8\ \Omega = 16 \text{ volts}$$
$$V_{R_1} = 16 \text{ volts}$$
$$V_{R_2} = 4 \text{ volts}$$
$$V_{R_3} = \frac{8}{3} \text{ volts}$$
$$V_{R_4} = \frac{4}{3} \text{ volts}$$

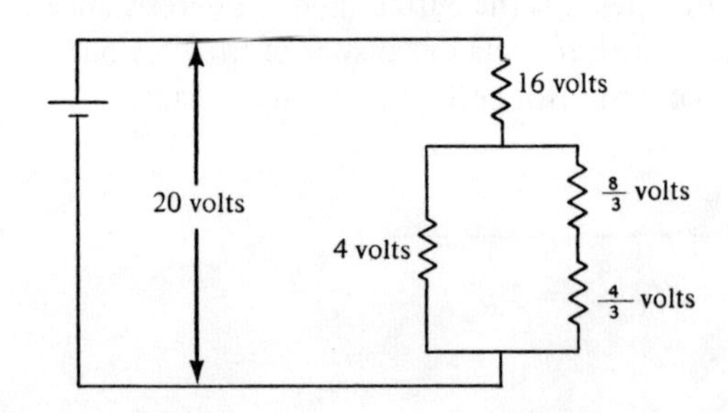

We can see here how Kirchhoff's voltage rule is satisfied.

Problem 3

Determine the current through each resistor.

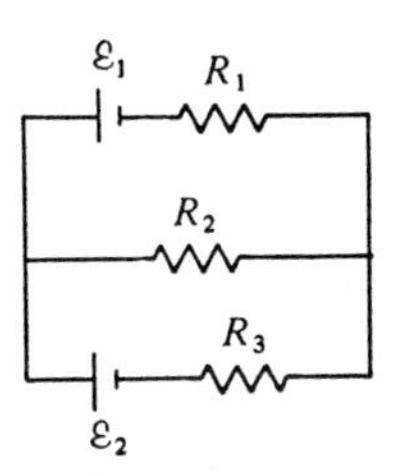

$\mathcal{E}_1 = 5$ volt
$\mathcal{E}_2 = 10$ volt
$R_1 = 2\ \Omega$
$R_2 = 4\ \Omega$
$R_3 = 3\ \Omega$

3.1 The specific objective of this problem is to provide practice in writing equations satisfying Kirchhoff's rules for current and voltage. These equations are most easily written when a given set of sign conventions are selected and applied consistently.

We start with a circuit less complicated than the problem. For the circuit shown here, the arrows indicate the conventional direction of positive current. Note that in the circuit external to the battery the current direction is from the positive to the negative terminal of the battery. This is a convention applicable to all problems.

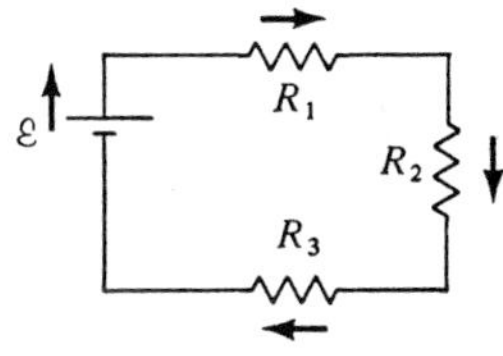

We want to write Kirchhoff's rule concerning voltages around a closed path as

$$\Sigma \mathcal{E} = \Sigma iR$$

The lefthand side is the sum of all emf's and the righthand side is the sum of all so-called iR drops. The rule for $\Sigma \mathcal{E}$ is that they are entered as positive if you establish a loop which goes from the negative to the positive terminal through the battery.

$\Sigma \mathcal{E}$ = ____________________

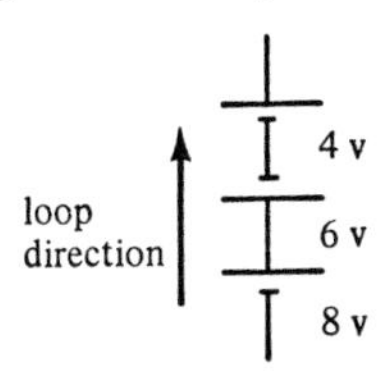

$\Sigma \mathcal{E} = 8 - 6 + 4 = 6$ volts (The loop direction proceeds from – to + for the 8 volt and 4 volt battery so they are entered as plus. The loop direction proceeds from + to – for the 6 volt battery so it is entered as negative.)

3.2 Try these configurations.

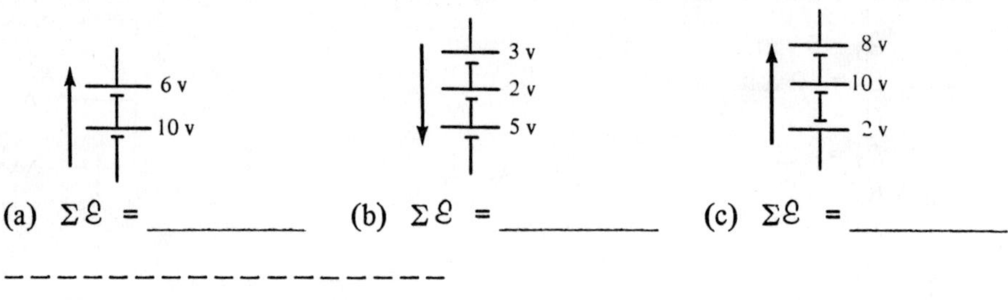

(a) $\Sigma\mathcal{E}$ = ________ (b) $\Sigma\mathcal{E}$ = ________ (c) $\Sigma\mathcal{E}$ = ________

(a) +16 v; (b) -10 v; (c) + 16 v

3.3 The convention for the iR drops is that they are entered as positive if the loop is traversed in the direction of an assumed current direction. Conversely they are negative if the loop is traversed in a loop opposite to the assumed current direction.

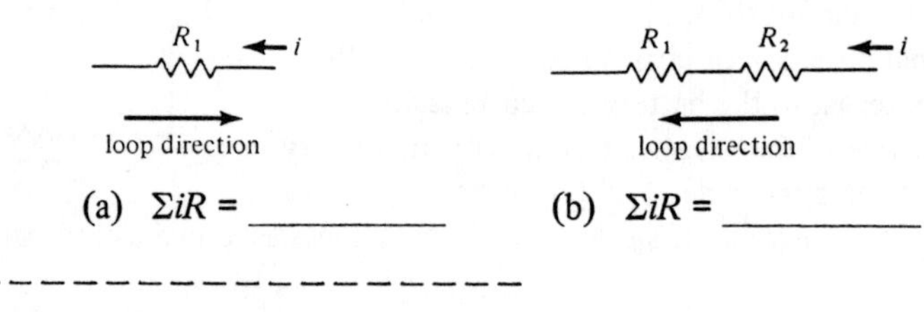

(a) ΣiR = ________ (b) ΣiR = ________

(a) $-i_1R_1$; (b) $+iR_1 + iR_2$

3.4 Write Kirchhoff's voltage rule for the circuit shown. The current directions are assumed. The loop direction is taken as clockwise.

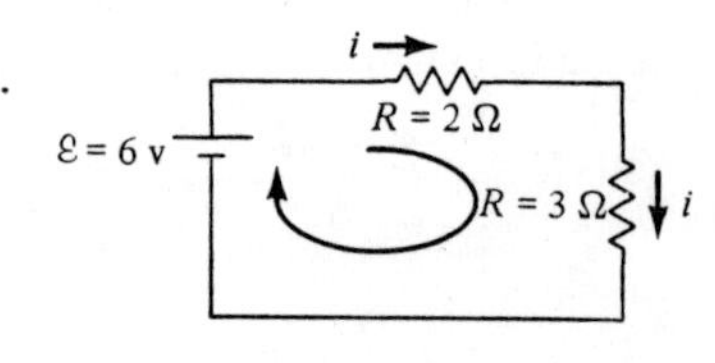

$6 \text{ v} = 2i + 3i = 5i$

3.5 For multiloop circuits we use the same conventions for the voltage equation. We guess at current directions, pick a certain loop, and write consistent equations.

There are three possible loops in this circuit. Identify them with loop arrows.

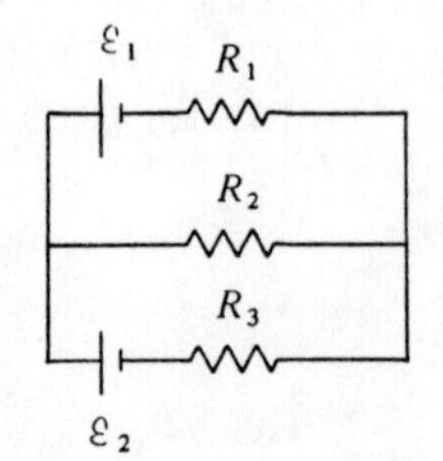

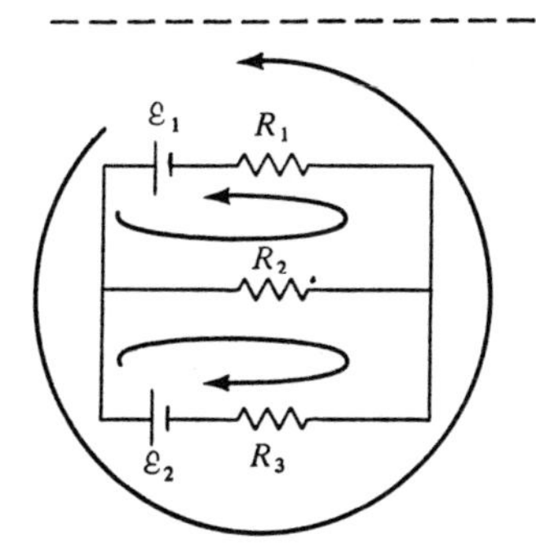

You may have your loops in opposite directions but as long as you are consistent your answer is correct.

3.6 It turns out that one does not have to use every possible loop. Each loop will result in an equation containing unknowns. Obviously you will need as many equations as there are unknowns.

In the diagram currents i_1, i_2, and i_3 are assumed as shown. Write the loop equation ($\Sigma\mathcal{E} = \Sigma iR$) for the upper loop.

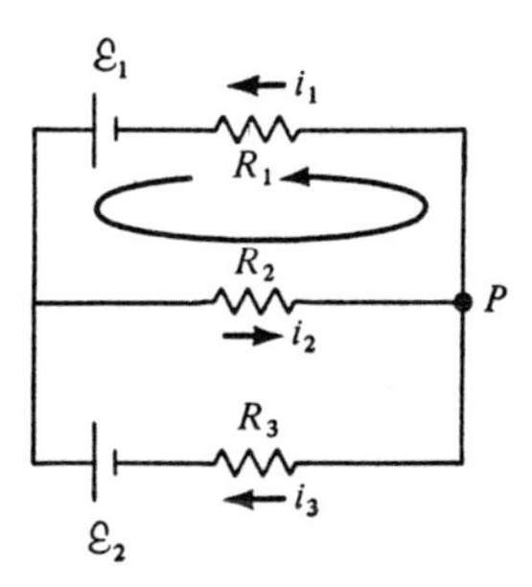

$\mathcal{E}_1 = i_1R_1 + i_2R_2$

$\mathcal{E}_1$ is positive because the loop goes from – to + through the battery; i_1R_1 and i_2R_2 are positive because the loop goes in the direction of the assumed current.

3.7 Pick a loop direction for the lower loop of the circuit in frame 3.6 and write the voltage equation.

For clockwise loop: $\mathcal{E}_2 = i_3R_3 + i_2R_2$
For counterclockwise loop: $-\mathcal{E}_2 = -i_3R_3 - i_2R_2$

3.8 If we put values from the problem into the loop equations from the answer to frames 3.6 and 3.7 we have

$$5 = 2i_1 + 4i_2$$
$$10 = 3i_3 + 4i_2$$

How many unknowns?

Three (i_1, i_2, i_3)

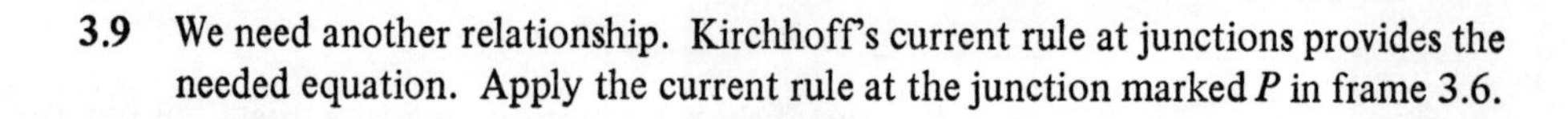

3.9 We need another relationship. Kirchhoff's current rule at junctions provides the needed equation. Apply the current rule at the junction marked P in frame 3.6.

$i_2 = i_1 + i_3$

$\Sigma i = 0 = i_2 - i_1 - i_3$
i_2 into junction is positive.
i_2 and i_3 from junction are negative.

3.10 The equations available are

$$\text{(a)} \quad 5 = 2i_1 + 4i_2$$
$$\text{(b)} \quad 10 = 3i_3 + 4i_2$$
$$\text{(c)} \quad i_2 = i_1 + i_3$$

We now have three equations with three unknowns. The physics of the problem is finished. The equations must now be combined. Solve (c) for i_3 and substitute into (b).

$10 = -3i_1 + 7i_2$

$i_3 = i_2 - i_1$
$10 = 3i_3 + 4i_2$
$10 = 3(i_2 - i_1) + 4i_2$
$10 = 7i_2 - 3i_1$

3.11 The previous answer can now be combined with equation (a) to eliminate either the i_1 or i_2.

$$\text{(a)} \quad 5 = 2i_1 + 4i_2$$
$$10 = -3i_1 + 7i_2$$

Eliminate i_1 and solve for i_2. Leave i_2 as a fraction.

$i_2 = \frac{35}{26}$ amp $= 1.3$ amp

$15 = 6i_1 + 12i_2$
$20 = -6i_1 + 14i_2$
$35 = 26i_2$
$1.3 \text{ amp} = i_2$

3.12 The fact that i_2 turned out positive means that the direction assumed for i_2 in frame 3.6 was correct. A negative solution (assuming no algebra errors) would indicate a wrong guess about direction.

Now substitute the numerical value of i_2 into equation (a) of frame 3.10 and solve for i_1.

$i_1 = -0.1$ amp (The minus sign means that the assumed direction of i_1 in frame 3.6 is backwards.)

3.13 You can now use equation (c) of frame 3.10 to determine the remaining unknown current. Do so.

---

i_3 = 1.4 amp (with the direction as first assumed)

$i_2 = i_1 + i_3$
1.3 amp = -0.1 amp + i_3

SELF–TEST

1 Determine the total resistance between points A and B in the circuit shown.

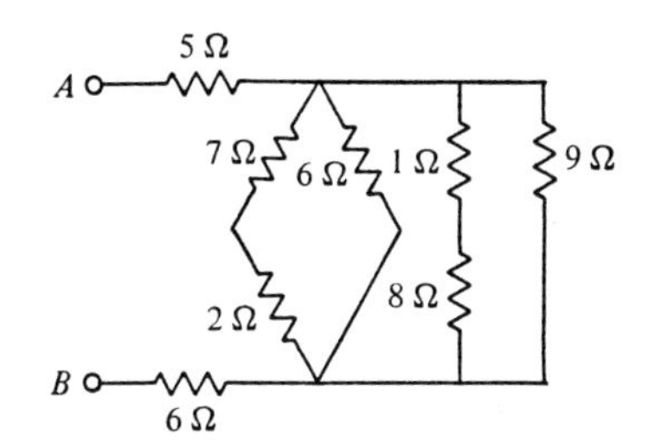

2 In the circuit shown the current through the 2 Ω resistor is 0.5 amp. Ignoring any internal resistance of the battery, calculate the emf of the battery.

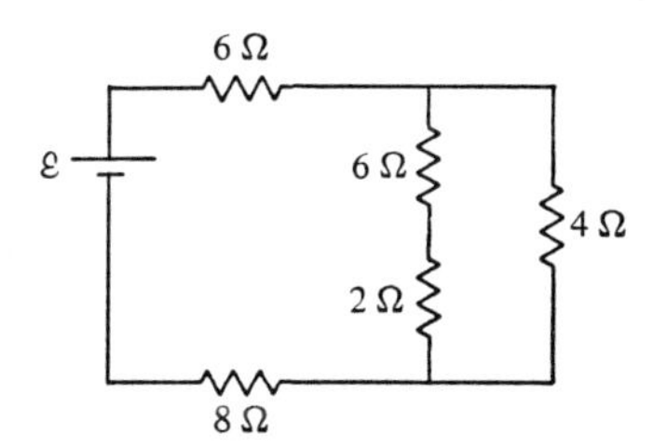

3 In the circuit shown determine the current in the 10 Ω resistor.

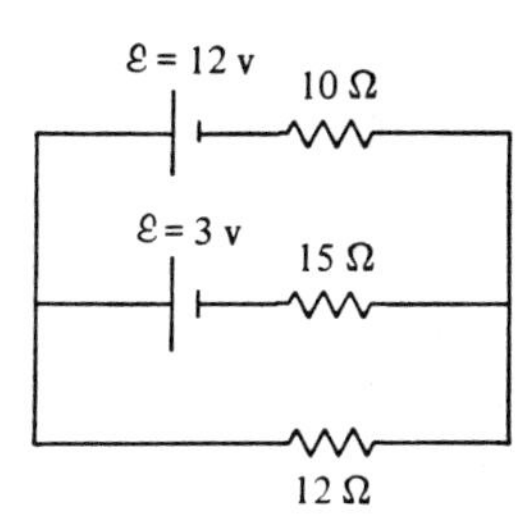

4 Each battery is 2 volts. Determine the voltage drop across the 2 Ω resistor.

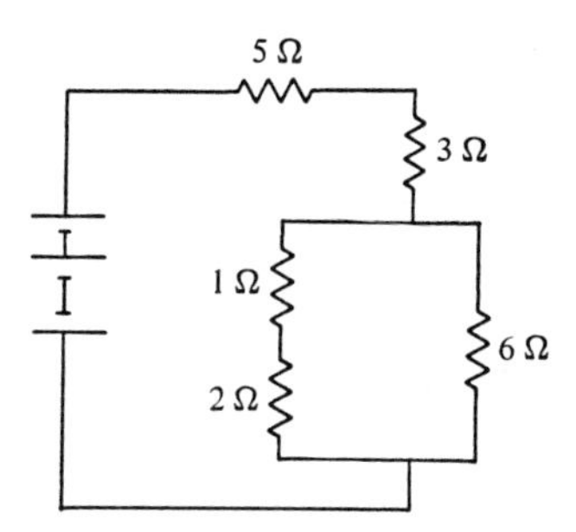

Answers to Self-Test

1 $13\ \Omega$

2 $\mathcal{E} = 25$ volts

3 $i = 0.64$ amp

4 $v = 0.27$ volts

CHAPTER SEVEN

Electric Energy, Heat, and Power

If the sample problems and objectives below identify your weak points, go directly to the programmed study section on page 118. If not, try the problems and compare your answers with those that follow. If you can do all the problems easily and if you are familiar with the objectives, you may wish to skip all or part of this chapter. The programmed study section covers techniques and concepts basic to solving the sample problems and fulfilling the objectives in this chapter. A programmed, step-by-step solution of each sample problem begins on page 121. A self-test is included at the end of the chapter.

SAMPLE PROBLEMS AND OBJECTIVES

Problem 1

A 60 watt household bulb operates on a line voltage of 120 volts. Calculate

(a) the resistance of this bulb when it is operating
(b) the current through the bulb when operating

Objectives:
1. Discussing the relationship between power, current, voltage, and resistance.
2. Discussing the heating effect of an electrical current.

Problem 2

An electric motor on a hoist requires 10 amp at 120 volts to lift 1000 lb at a rate of 15 ft/min. How efficient is this crane?

Objectives:
1. Discussing the relationship between electrical and mechanical energy.
2. Defining efficiency.
3. Discussing conversion factors.

Problem 3

Many homes use electric heating units for heat. For a unit rated at 1100 watts, 220 volts, what is the operating resistance of the heater, the current through the heater, and the cost per hour at 4¢ per kilowatt-hour?

Objective: 1. Practical problem

Answers to Sample Problems

See page 121 for programmed, step-by-step solutions to these problems.

Problem 1

$R = 240$ ohm
$i = 0.5$ amp

Problem 2

e = 28%

Problem 3

$R = 44\ \Omega$
$i = 5$ amp
Cost = 4.4¢ per hour

PROGRAMMED STUDY SECTION

1 As in the chapter on current electricity we will review the concepts of electric energy, heat, and power by discussing light bulbs. What electrical information is stamped on ordinary household light bulbs?

The power rating (e.g., 100 watts) and the voltage required (e.g., 120 volts)

2 If you add another 100 watt bulb in parallel, what changes occur in the circuit?

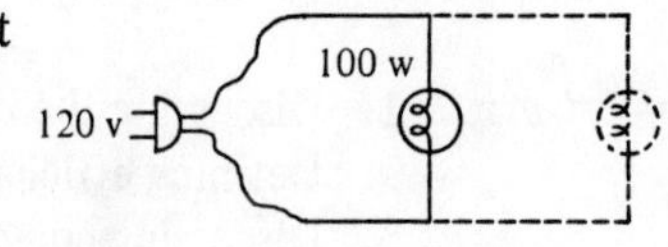

The total power delivered will double (there are now two 100 watt bulbs) and the total current from the 120 volt source will increase.

3 Will the current through each light bulb be the same?

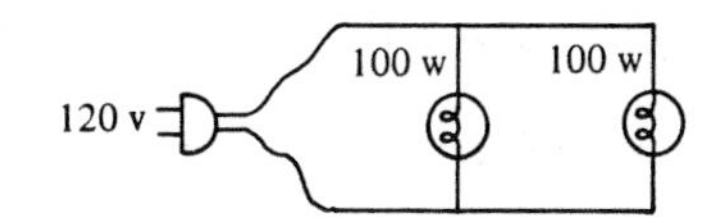

Yes (It makes sense to assume that two identical bulbs would have the same current since they are each connected to 120 volts.)

4 The point of the previous two frames is that as one adds identical bulbs in parallel the total power increases and the total current in the circuit increases.

In the graph to the right one data point is plotted correlating the power delivered and the corresponding current for one bulb. Add data points for a second, third, fourth, and fifth bulb connected in parallel.

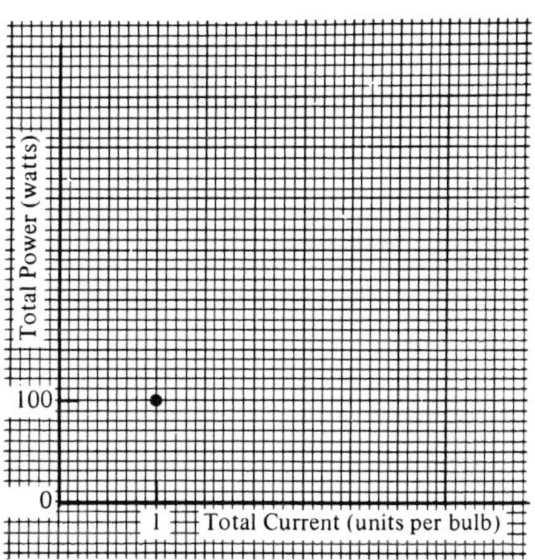

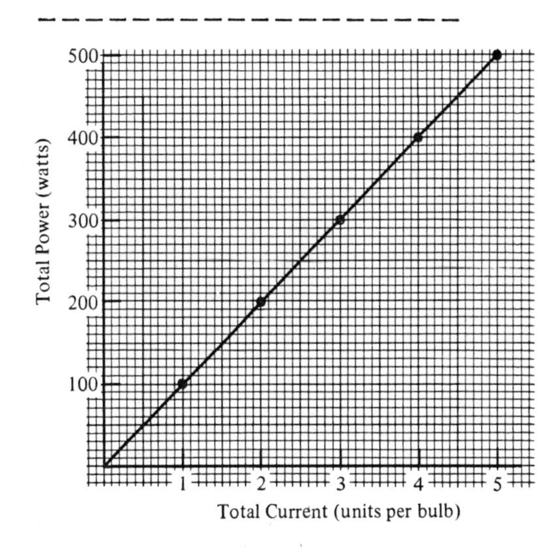

Example: For 3 bulbs the total power is 3 X 100 watts and the total current is 3 X 1 unit of current.

5 The variables power and current represented in the previous answer fit well to a line of constant slope. As usual for graphs of this type

$$P = \text{constant} \times i$$

Let's give a physical interpretation to the constant. What is there about the circuit which doesn't change as bulbs are added in parallel?

The constant is the outlet voltage which is always 120 volts.

6 The result

$$P = iV$$

is the electrical power transferred to an element. Using Ohm's law we can write the equation in other ways for resistive devices.

$$P = iV = \frac{V^2}{R} = i^2R$$

The unit of power is the joule/sec or watt.

7 From the definition of power as

$$P = iV$$

where P has the units joule/sec, what are the units of iVt, where t is the elapsed time?

---

joule (This is a unit of energy.)

8 We have then

$$W = iVt$$

as a statement of electrical energy having the units of energy.

In many problems in physics we need to relate electrical energy to other kinds of energy (e.g., mechanical and thermal). The frequent task in such problems is to recognize that

$$W_{\text{elec}} = kW_{\text{mech, etc.}}$$

where the proportionality constant k is a conversion factor. For example, 1 joule = 0.239 calorie.

SOLUTIONS TO SAMPLE PROBLEMS

Problem 1

A 60 watt household bulb operates on a line voltage of 120 volts. Calculate

(a) the resistance of this bulb when it is operating
(b) the current through the bulb when operating

1.1 Very nearly all of the electrical power delivered to an incandescent light bulb is dissipated in the form of heat. This heat is confined to the filament of the bulb and the filament is designed for that purpose. The power delivered to the filament is the product of the potential difference across the filament and the current through the filament.

In this problem $i =$ ________________ amp through the filament.

$i = 0.5$ amp

$$P = iV$$
$$60 \text{ watts} = i \times 120 \text{ volts}$$

1.2 Light bulbs do not qualify as ohmic devices because the resistance of the filament is not constant. The resistance depends on temperature. This means that Ohm's law, $V = iR$, is not valid for all values of V and i. However, the operating resistance of an electrical device can always be determined by specific instances of current and voltage.

What is the resistance of the bulb when it is turned on and operating at 120 volts with a current of 0.5 amperes?

$R = 240\ \Omega$

$$R = \frac{V}{i} = \frac{120 \text{ volt}}{0.5 \text{ amp}}$$

1.3 The value of resistance can be thought of as the operating resistance. When do bulbs usually burn out—when they are turned on or when they are already on?

Usually they pop when you first throw the switch.

1.4 The reason for the previous answer is that when the filament is cold (the bulb is off) the filament resistance is small (say, much less than 240 Ω). When first turned on, how would the current in the filament compare to the answer of frame 1.2?

It would be larger than 0.5 amp. This large initial current is responsible for destroying the filament when the bulb actually burns out.

Problem 2

An electric motor on a hoist requires 10 amp at 120 volts to lift 1000 lb at a rate of 15 ft/min. How efficient is this crane?

2.1 As in problem 1 we will first calculate the power delivered to the hoist motor.

$P =$ ____________ watts

$P = 1200$ watts

2.2 Convert the previous answer to kilowatts.

$P = 1.2$ kw (1 kilowatt = 1000 watts)

2.3 Now we can turn our attention to the movement of the load. We will assume that the load is always moving with the fixed speed of 15 ft/min. What does such an assumption reveal concerning the net force on the load? (Dredge up your understanding of dynamics.)

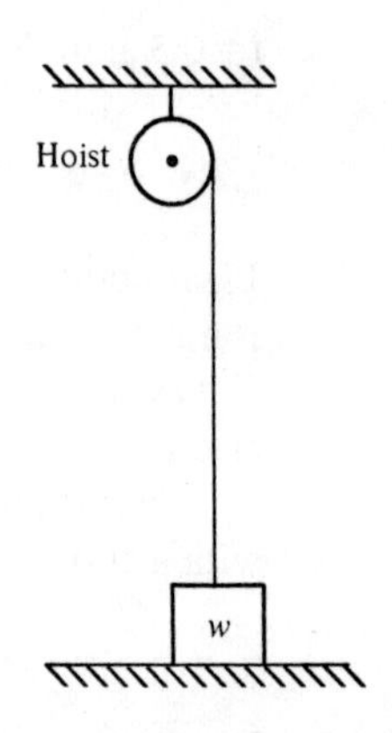

Fixed speed means no acceleration, thus the net force is zero.

2.4 Given the previous answer, what is the tension in the hoist cable attached to the 1000 lb load?

1000 lb

T
w
w

T = **w** in magnitude, but of course oppositely directed.

2.5 A fixed force of 1000 lb is moving the load at 15 ft/min. How many ft will the load move in 20 sec?

---

5 ft

$20 \text{ sec} = \frac{1}{3} \text{ min}$
$15 \text{ ft/min} \times \frac{1}{3} \text{ min} = 5 \text{ ft}$

2.6 In physics we define the work done by a force as

$$W = \mathbf{F} \cdot \mathbf{s}$$

How much work is done by the crane in 20 sec?

$W = 5000$ ft-lb

$W = 1000 \text{ lb} \times 5 \text{ ft} \times \cos 0^\circ$
$W = 5000$ ft-lb

2.7 In an earlier frame we calculated that the hoist motor requires 1200 watts while operating continuously. A watt is a joule/sec. How much energy in joules is required by the hoist motor during the 20 sec that it is doing 5000 ft-lb of work?

24,000 joule

$W = iVt$
$W = 10 \text{ amp} \times 120 \text{ volt} \times 20 \text{ sec}$
$W = 1200 \text{ joule/sec} \times 20 \text{ sec}$
$W = 24{,}000$ joule

2.8 So far we have found that 24,000 joule of electrical energy is delivered to the hoist motor so it can do 5000 ft-lb of work. These are not the units we normally use to describe mechanisms like motors. The power of electrical motors, automobiles, etc., is usually characterized in units of ________________ power.

---

horsepower

2.9 The term horsepower really means the work per unit time delivered by a motor, a car, or even a horse. In physics we express work in three different units.

(a) ____________ for English units

(b) ____________ for MKS units

(c) ____________ for cgs units

(a) ft-lb; (b) joule; (c) ergs

2.10 The work per unit time of the hoist is

$$\frac{W}{t} = \frac{5000 \text{ ft-lb}}{20 \text{ sec}} = 250 \text{ ft-lb/sec}$$

This is a constant. By definition 1 hp = 550 ft-lb/sec. How many hp is actually required to lift the load through 15 ft?

$\frac{5}{11}$ hp

2.11 For electrical energy

$$1 \text{ kw} = 1\tfrac{1}{3} \text{ hp}$$

Out hoist required 1.2 kw of electrical power to deliver $\frac{5}{11}$ hp the *output.* What is the horsepower delivered to the hoist motor (which we will call the *input*)? Express the answer as a whole number and a fraction.

$1\frac{3}{5}$ hp

$$1.2 \text{ kw} \times \frac{1\frac{1}{3} \text{ hp}}{\text{kw}}$$

2.12 Finally we define efficiency to be

$$e = \frac{\text{hp output}}{\text{hp input}}$$

$e =$ ________________

---

about 28%

$$e = \frac{\frac{5}{11}}{1\frac{3}{5}} = \frac{25}{88}$$

2.13 Before leaving this problem it may be useful to review the ideas. For the load

$$\text{power output} = Fv$$
$$\text{power input} = iv$$

These have different units. It was necessary to convert them both to hp or some other suitable unit. We used

$$1 \text{ kw} = 1\tfrac{1}{3} \text{ hp}$$
$$1 \text{ hp} = 550 \text{ ft-lb/sec}$$

Problem 3

Many homes use electric heating units for heat. For a unit rated at 1100 watts, 220 volts, what is the operating resistance of the heater, the current through the heater, and the cost per hour at 4¢ per kilowatt-hour?

3.1 This problem is essentially like problem 1 with the added question about cost. What current is required to deliver 1100 watts at 220 volts?

$i = 5$ amp

$$\frac{P}{V} = \frac{1100 \text{ watts}}{220 \text{ volts}} = 5 \text{ amp}$$

3.2 What is the operating resistance of the heater?

$R = 44\ \Omega$

$$R = \frac{V}{i} = \frac{220 \text{ volts}}{5 \text{ amps}} = 44\ \Omega$$

3.3 Now we need consider the cost involved. The problem gives the cost as 4¢ per kilowatt-hour. Let us try to establish some physical feeling for what is going on rather than only manipulating formulas.

In the chapter on electric potential we defined potential difference as the work per unit charge necessary to transfer charge from one point to another in a circuit.

$$\Delta V = \frac{W}{q}$$

Calling the potential difference across the heating element V we have

$$W = qV$$

Generally it makes no difference whether the heating element is large enough to heat a room or just large enough to heat a cup of tea water. V is the same if each is plugged to the same outlet.

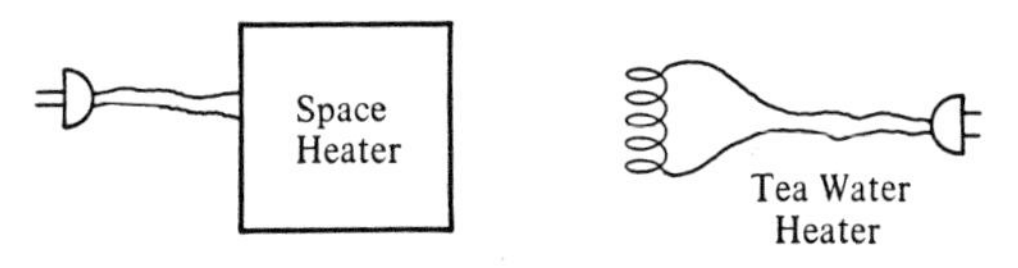

Which of the above is more expensive to operate?

Clearly the space heater—even one that plugs into a 120 volt outlet.

3.4 Here we explicitly show V the same for each heater. We use a battery for clarity.

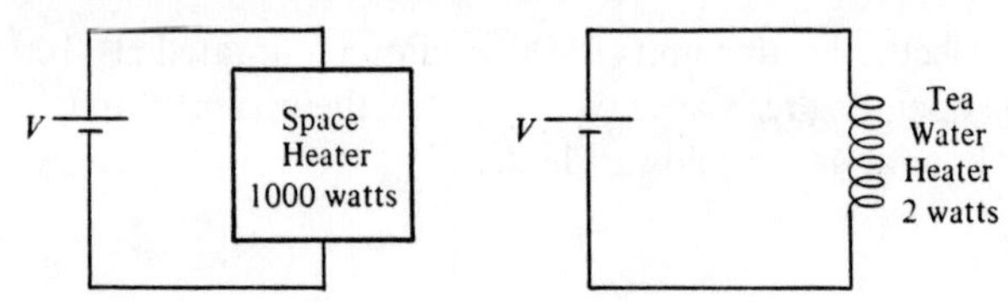

Each charge moving in either circuit will decrease its electrical potential energy by qV, thus delivering energy to the space heater or water heater. Now surely the charges constituting the electric current are the same in both cases and we explicitly impose the condition that V is the same. How is it then that over equal intervals of time the charges in the space heater circuit deliver more energy than the charges in the water heater?

There must be more charges per unit time in the space heater circuit. For example, in 1 min 10 charges go through the space heater (each delivering qV energy) while only 1 charge goes through the tea water heater (also delivering qV energy).

3.5 What is the electrical concept which relates the number of charges moving through a circuit element in a given time?

---

current ($i = \Delta q/\Delta t$)

3.6 What electrical characteristic permits i of the space heater to be larger than i of the tea water heater?

R (The space heater must have a smaller resistance: $i = V/R$.)

3.7 Starting with the idea of work ($W = qV$) we see that it is more useful to use

$$\frac{\Delta W}{\Delta t} = \frac{\Delta q}{\Delta t} V$$

$$P = iV$$

This includes explicitly the current which differentiates the power delivered to different objects connect to the same outlet. In your home all objects are either 120 or 240 volts, so V isn't what the electric company keeps track of in order to bill you. In terms of units

$$1 \text{ watt} = 1 \text{ joule/sec} = 1 \text{ coul/sec} \times 1 \text{ joule/coul}$$

$$P = i \times V$$

Similarly

$$1 \text{ kw} = 1000 \text{ joule/sec}$$

In this problem the rate of energy delivery is 1100 joule/sec.

$P =$ ____________kw

1.1 kw

3.8 At a cost of 4¢ per kilowatt-hour, what is the cost of this heater per hour?

4.4¢ per hour

SELF–TEST

1 Assume that the operating resistance of a light bulb does not change with temperature. Show quantitatively that a 100 watt, 120 volt household bulb has less resistance than a similar 25 watt bulb. Also determine quantitatively which of these two bulbs would light if they were connected in series across 120 volts. (Hint: Determine the operating current for each individually.)

2 A toaster requires 1600 watts and operates 30 minutes at breakfast time. What is the cost per 30-day month at 5¢ per kilowatt-hour?

3 During a "brown-out" the line voltage in your home drops to 80 percent of its normal value of 120 volts. Assuming the resistance of a 1000 watt heating element to be constant, what power is delivered to the heater during the brown-out?

4 A current of 3 amp passes through a 25 Ω resistor for 1 hour. If the potential difference is 75 volts, how many coulombs pass through the resistor and what is the total energy delivered by the battery.

Answers to Self-Test

1 R of the 100 watt bulb is 144 Ω
R of the 25 watt bulb is 580 Ω
In series the current through each is 0.17 amp which is nearly the operating current of the 25 watt bulb. The 100 watt bulb requires a current of 0.83 amp. Only the 25 watt bulb would light in a series configuration.

2 $1.20 per month.

3 640 watts

4 10,800 coul, 8.1×10^5 joule

CHAPTER EIGHT
Reflection and Refraction

If the sample problems and objectives below identify your weak points, go directly to the programmed study section on page 129. If not, try the problems and compare your answers with those that follow. If you can do all the problems easily and if you are familiar with the objectives, you may wish to skip all or part of this chapter. The programmed study section covers techniques and concepts basic to solving the sample problems and fulfilling the objectives of this chapter. A programmed, step-by-step solution of each sample problem begins on page 142. A self-test is included at the end of the chapter.

SAMPLE PROBLEMS AND OBJECTIVES

Problem 1

Calculate the critical angle which results in total reflection at the boundary of glass and air.

Objectives:

1. Clarifying incident and refracted rays as well as incident and refraction angles.
2. Discussing Snell's law including ray diagrams.

Problem 2

Describe fully the images formed by a convex lens of focal length 10 cm when the object is placed first 20 cm away from the lens and then 5 cm away.

Objectives:

1. Discussing ray diagrams with particular emphasis on those rays useful in determining images, both real and virtual.
2. Discussing sign conventions and their interpretation.
3. Geometric interpretation of magnification.

Problem 3

A dental mirror of radius of curvature 2 in. is held $\frac{3}{4}$ in. from a tooth. Describe the character, size, and position of the image formed. Should the mirror be plane, concave, or convex?

Objectives:
1. Discussion of ray diagrams for mirrors.
2. Comparing ray diagrams with the sensible use of a mirror.

Answers to Sample Problems

See page 142 for programmed, step-by-step solutions to these problems.

Problem 1

$\theta = 42°$ in going from glass to air

Problem 2

For object outside the focal length the image is real, inverted, 20 cm from the lens, and the same size as the object. For object inside the focal length the image is virtual, erect, -10 cm from the lens, and magnified by a factor of 2.

Problem 3

$m = +4, q = -3$ in.
The image is a virtual erect image.
The mirror is concave.

PROGRAMMED STUDY SECTION

1 The study of geometrical optics is aptly named. Successful problem solving in optics is strongly dependent on geometry which includes things like measuring angles, drawing neat diagrams, and comparing similar triangles. We will concentrate on geometry in these programmed study frames.

One of the premises of geometric optics is that a light ray travels in a straight line within a given medium. The circles below contain water. The lines represent light rays passing from air into water (i.e., two different media). Which sketch below does not satisfy the premise concerning a straight line ray in a given medium?

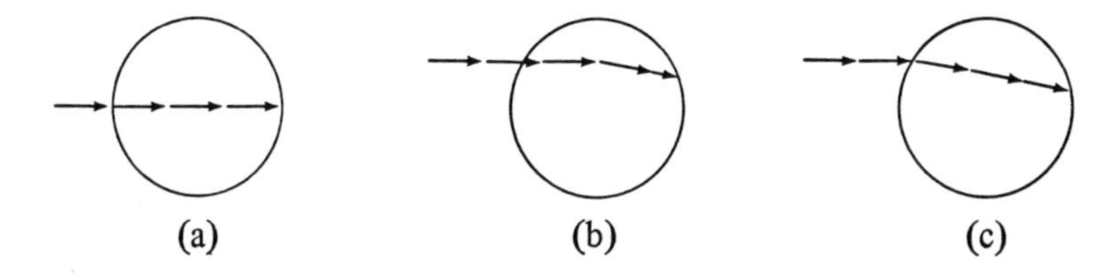

(b) The ray in the water consists of two straight lines. The premise is that a ray in a medium travels in one straight line.

2 Do the three rays below violate the straight line premise? (The rectangle is glass.)

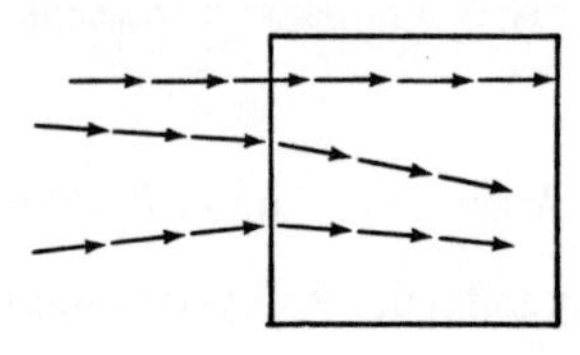

No. The rays travel in straight lines in both air and glass. The direction of the line may change in going from air to glass. The point so far is that there is only one straight line for a given ray in a medium. Of course, we are considering media which have constant optical properties.

3 Another premise of ray optics is that a ray may change direction as the ray goes from one medium to a different medium. Snell's law

$$n_i \sin \theta_i = n_r \sin \theta_r$$

describes the manner in which the straight line may change from one medium to another. The angles θ_i and θ_r are angles formed by a light ray and the perpendicular at the boundary between the different media.

Draw the appropriate incident angles for the rays shown. (Note that these rays also satisfy the first premise.)

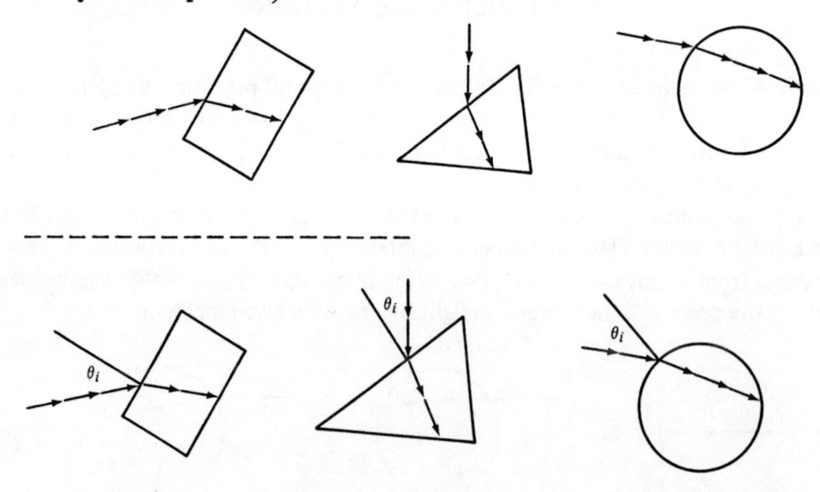

4 Snell's law $n_i \sin \theta_i = n_r \sin \theta_r$ describes the change of direction. In words, the equation can be though of as

n_i	$\sin \theta_i$	=	n_r	$\sin \theta_r$
medium of incident ray	direction of incident ray		medium of refracted ray	direction of refracted ray

The n is the index of refraction of the media in which the incident or refracted ray appears. For example, $n = 1.33$ for water, $n = 1$ for air, $n = 1.5$ for glass.

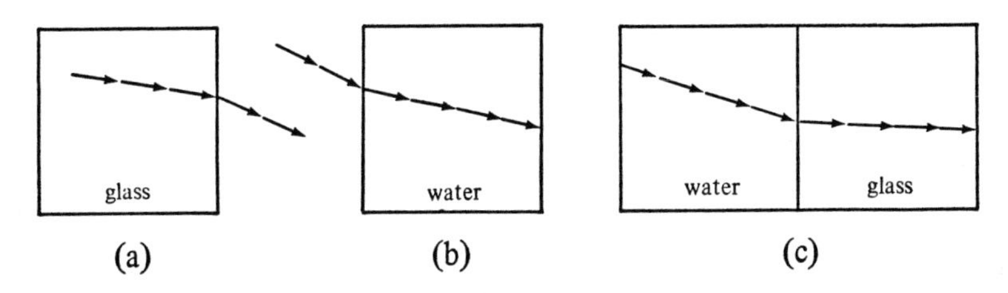

In each case, numerically identify n_r as it would appear in Snell's law.

(a) $n = 1$, the refracted ray is in air.
(b) $n = 1.33$, the refracted ray is in water.
(c) $n = 1.5$, the refracted ray is in glass.

5 Identify the incident angles θ_L on each sketch in frame 4.

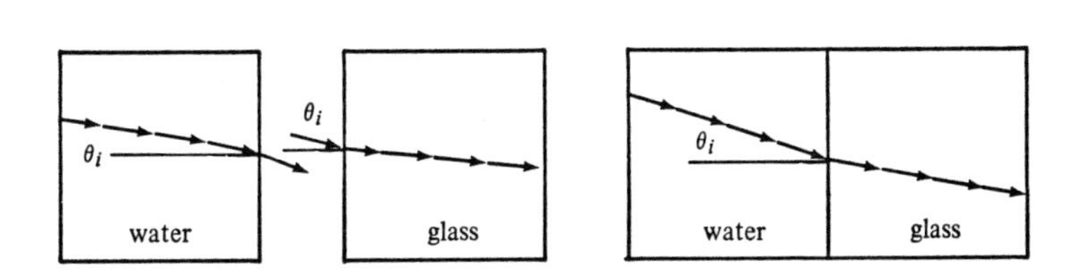

In each case θ_i is the angle formed by the incident ray and a perpendicular to the boundary where the medium changes.

6 The essential features of converging and diverging lenses can be understood by resorting to a close look at the requirements of Snell's law.

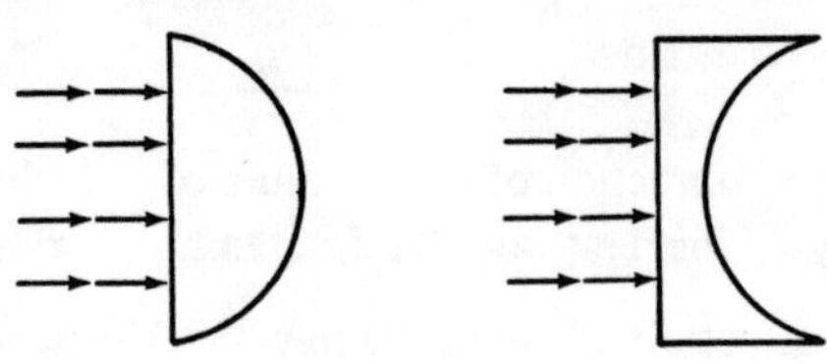

What are the incident angles for the rays shown incident on the flat side of the glass?

Zero

7 What does Snell's law

$$n_i \sin \theta_i = n_r \sin \theta_r$$
$$(1)(\sin \theta_i) = (1.5)(\sin \theta_r)$$

predict for $\sin \theta_r$ when $\theta_i = 0$ as in the previous frame?

Zero; therefore $\theta_r = 0$.

8 We now show a ray in glass.

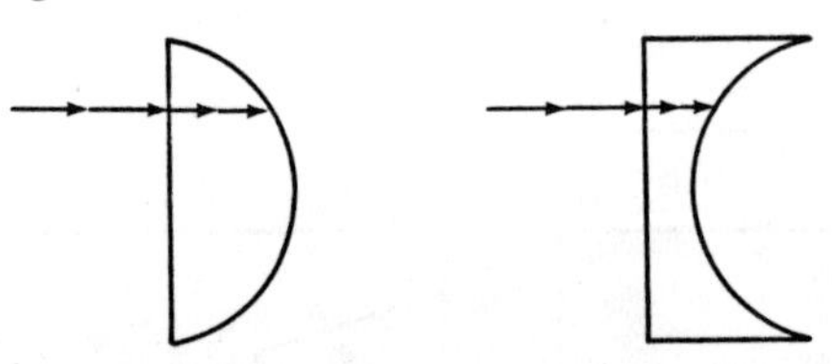

As the ray passes from glass back to air, Snell's law is

$$1.5 \sin \theta_i = 1 \sin \theta_r$$

Schematically identify θ_i in each case above.

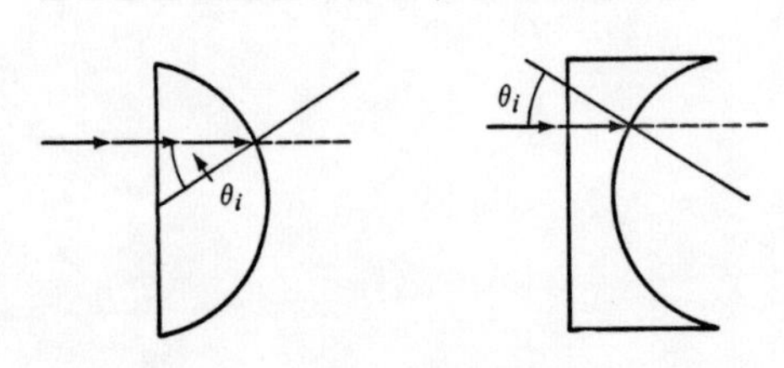

We add to the answers above a dotted line to represent the ray *if* the medium didn't change.

9 Now θ_i is the same in both cases because the curved portions have the same radius. What does the equation $1.5 \sin \theta_i = 1 \sin \theta_r$ require about the comparative size of θ_i and θ_r in going from glass to air?

$\theta_r > \theta_i$ in both cases

Note: (1.5) (number) = (1) (other number). Clearly the equality requires number < other number, or $\sin \theta_i < \sin \theta_r$, which means $\theta_i < \theta_r$.

10 Show qualitatively the direction of the refracted ray in both cases in the answer to frame 8.

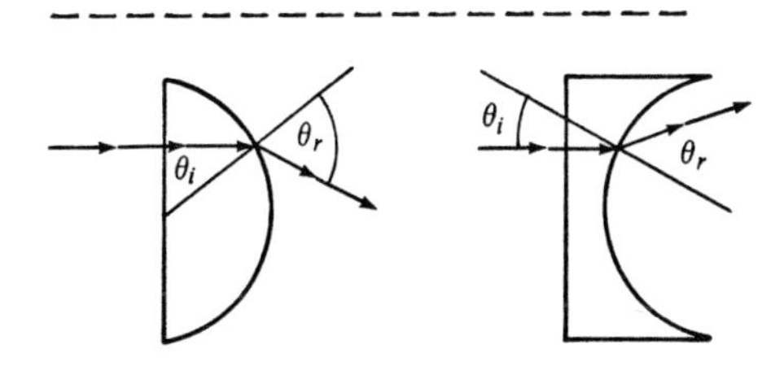

In both cases $\theta_r > \theta_i$ as required by Snell's law.

11 If we add other rays, the result is as sketched below.

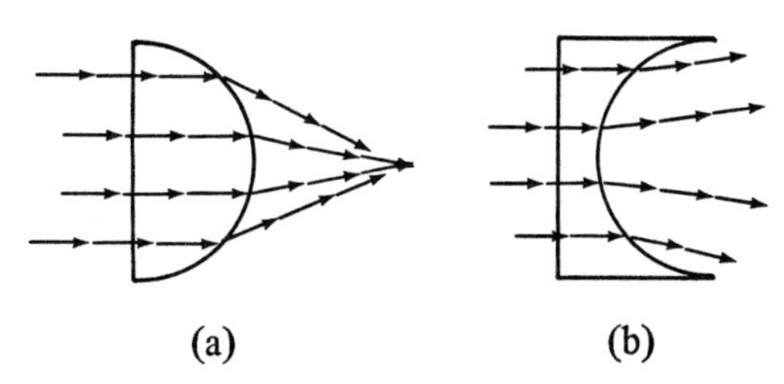

(a) (b)

In one case the lens converges the rays while in the other the lens diverges the rays. The effect of a lens on parallel rays gives rise to the concept of a point of focus. Which lens above brings the light rays to a point of focus?

(a)

12 Lenses which converge light rays are *convex.* Lenses which diverge light rays are *concave.* (Here we assume that the lens is optically denser than its environment.) What kind of lens is shown in (a) of frame 11?

Convex

13 In the next few frames we will develop the lens equation for a double convex lens. We will do so by considering the action of the lens on specific light rays.

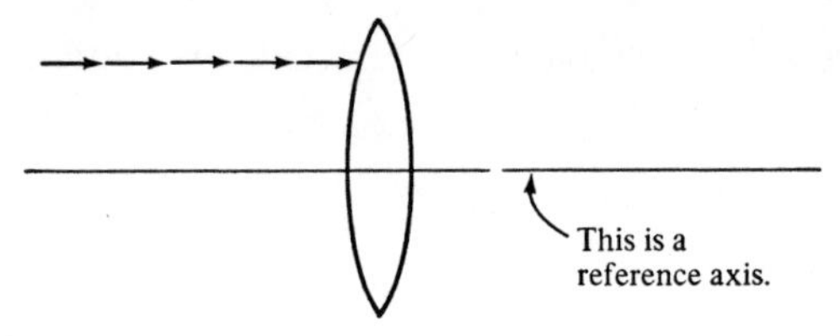

In the sketch, a ray parallel to the axis of the lens is incident on the lens from the left. Qualitatively, show the ray as it passes through the lens and into air again.

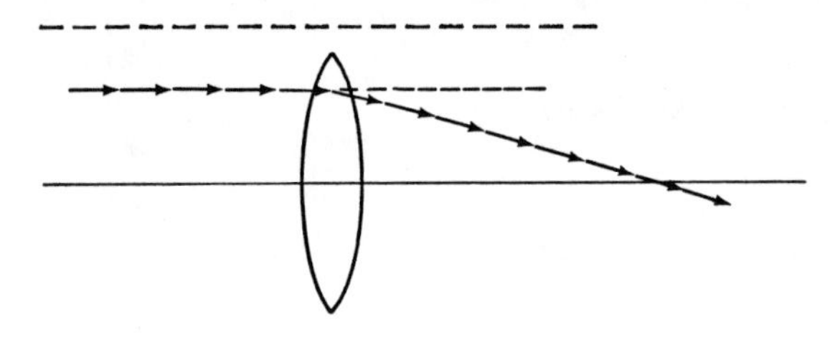

Note that the ray is refracted as it goes from air to glass and is again refracted as it goes from glass to air. The dotted line represents the incident ray if it were not refracted.

14 If we add other parallel rays incident from the left they would all meet at the point called the focal point of the lens.

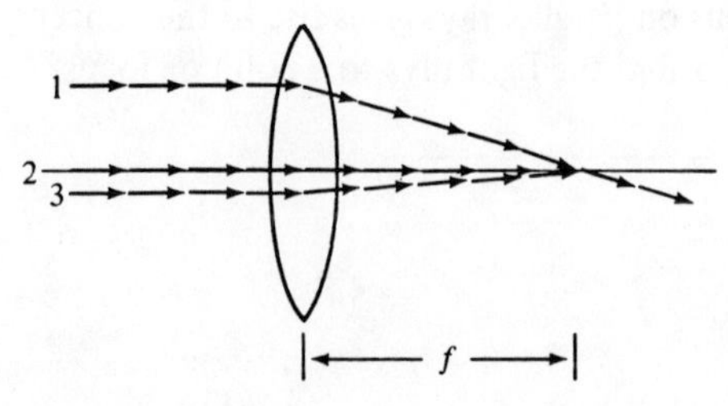

The distance from the center of the lens to the crossing point of incident parallel rays is called the *focal length.* Ray 2 is incident along the axis of the lens. Why does this ray not change direction when in fact it goes from air to glass to air again?

The incident angle for ray 2 is zero (i.e., the ray comes in along the perpendicular to the boundary between air and glass). According to Snell's law, such a ray has no refraction angle. The ray proceeds in a straight line.

15 Ray 2 in the previous frame is of special significance in solving problems dealing with thin lenses. Ray 2 passes through the center of the lens and is undeflected. Apply this principle to the three incident rays shown below.

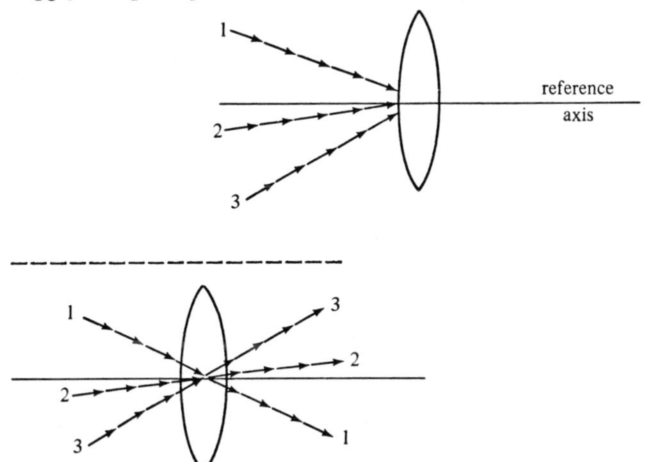

All three rays pass through the center and are not deflected from their incident direction. These rays are all incident on the lens in the region of the lens where the faces are essentially parallel. (See frames 2.2–2.4 in sample problem 2 for further discussion of this point.)

16 We need only two rays to discuss many of the features of geometric optics dealing with lenses. The two rays are:

- An incident ray parallel to the reference axis passes through the focal point on the reference axis on the opposite side of the double convex lens.
- An incident ray passing through the center of a thin lens is not deflected from its initial direction.

We will use these two rays to interpret the lens equation.

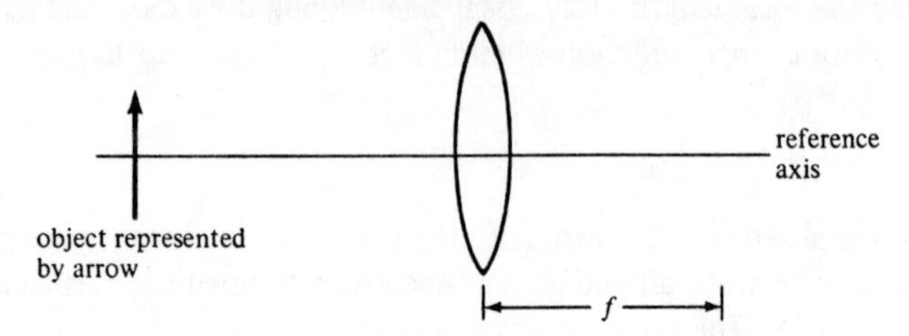

Carefully draw two rays which both originate from the top of the arrow. Indicate the rays in the lens as well as both sides of the lens.

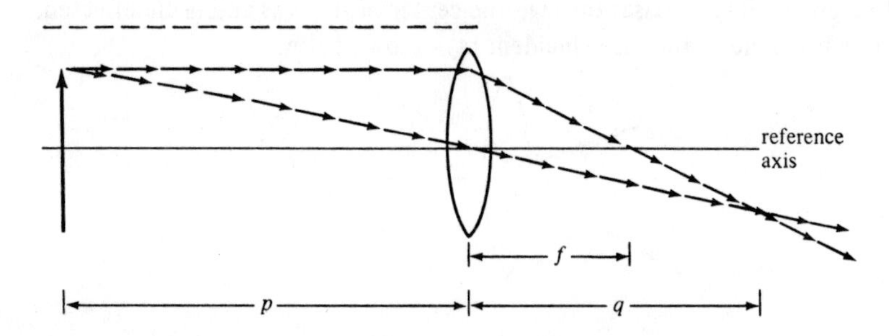

Obviously we pick two rays whose direction we know as alluded to in frame 16.

17 We add to the previous answer the distances p and q as shown.

> p = distance of object which is the source of light rays, as measured from the center of the lens
>
> q = distance of image of the object as measured from the center of the lens

We see in the previous answer that the light rays which *both* originated at the tip of the arrow again *come together* a distance q from the lens. This situation results in a real image being formed at q. In other words, all light rays from the tip of the arrow come together in space a distance q away from the lens and would reconstruct the object on a screen placed at q.

A geometric analysis (which is probably included in your textbook) would show that

$$\frac{1}{p} + \frac{1}{q} = \frac{1}{f}$$

This is the thin lens equation. (See sample problem 2 for an application.)

18 We will now consider mirrors and reflection. The only additional fact we need is the relationship between incident angles and reflected angles.

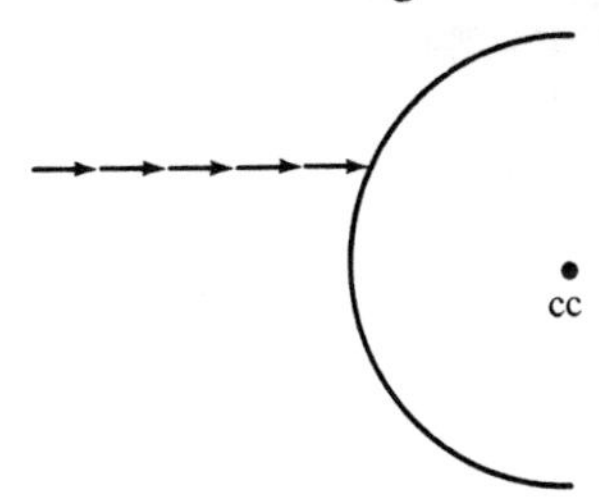

A ray is incident on a convex mirror surface as shown. The center of curvature of the mirror (cc) is indicated. Incident angles for mirrors are defined in the same manner as incident angles for refraction. Schematically show the incident angle.

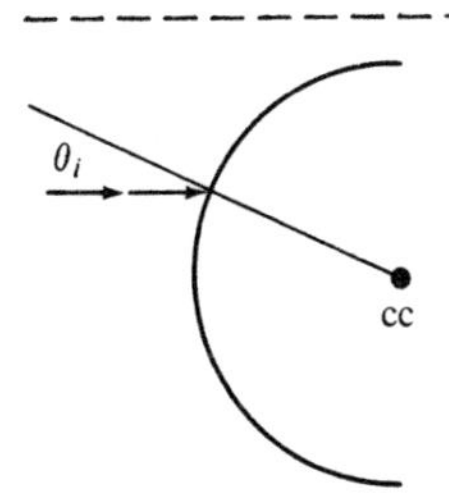

The incident angle θ_i is the angle between the incident ray and a perpendicular at the mirror surface. Note that the perpendicular is an extension of the radius of the spherical surface.

19 For reflection

$$\theta_i = \theta_r$$

where θ_r is the angle between the reflected ray and the perpendicular at the mirror. Schematically indicate the reflected ray in the answer frame above.

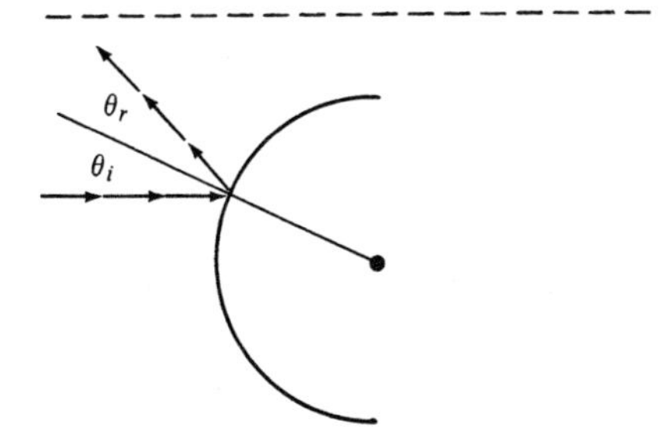

20 As in the case with lenses we are interested in the converging and diverging properties of mirrors. We are particularly interested when the incident rays start from the same point since that is how one constructs images.

Show by ray diagrams which of the mirrors below would converge the two incident rays shown. Continue the rays properly until they strike the mirrors.

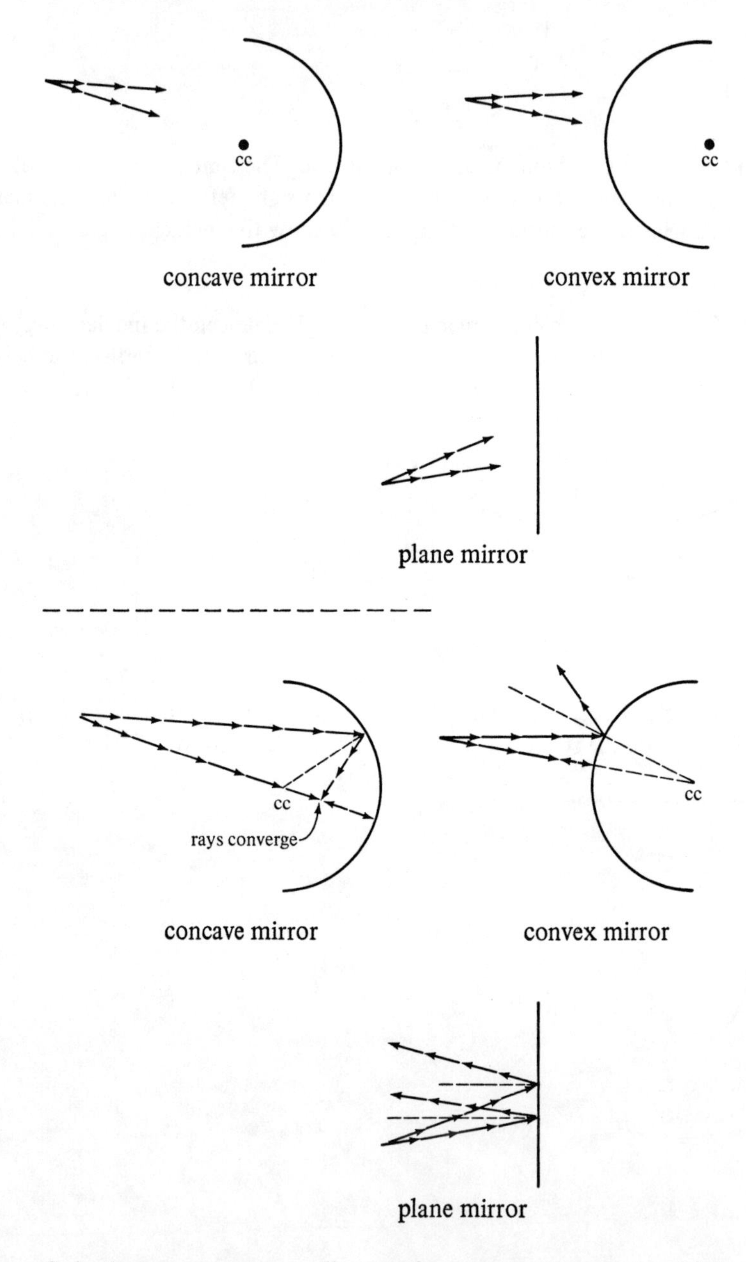

Only the concave mirror will cause the rays to converge at a point.

21 Perhaps you had some difficulty with the previous frame. If you did, it was probably with the incident ray which passed through the center of curvature (cc) before striking the mirror. Such rays have zero incident angle (they come in along the perpendicular); therefore, they reflect back along the original direction. Such a ray is a good one to remember for curved mirrors. It just goes back along the way it came in. It is analagous to the ray which passes through the center of a lens undeflected.

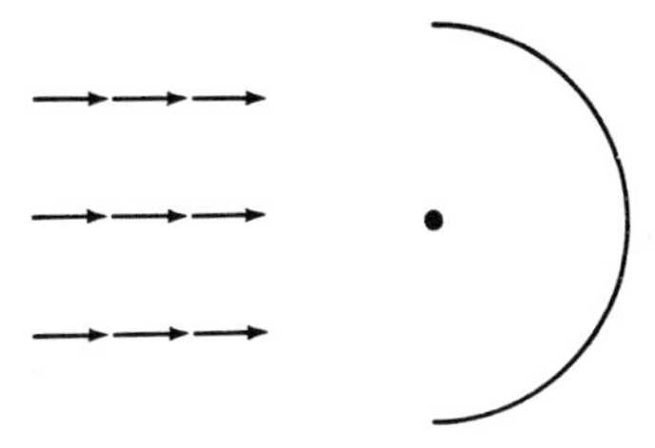

Do these three rays converge at a point? Determine the answer by a ray diagram.

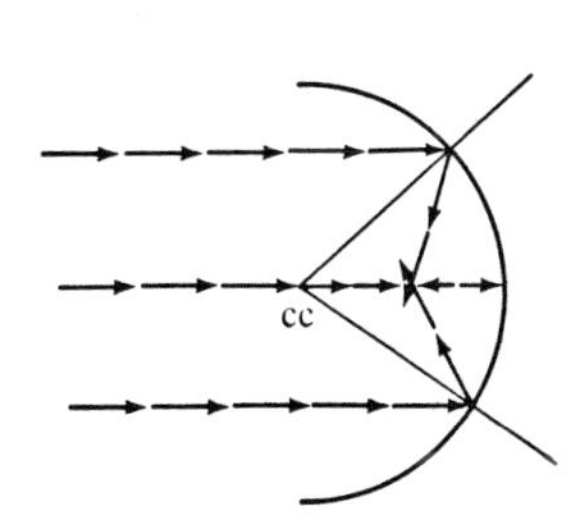

22 Again we define the point at which parallel rays converge as the focal point of the mirror.

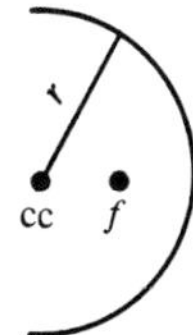

For mirrors

$$2f = r$$

where r is the radius of curvature.

What is the radius of curvature of a plane mirror?

∞ (infinite) A circle would have to have an infinite radius before its circumference would be straight.

23 What is the focal length f of a plane mirror?

∞ (infinite)

24 In the remaining frames we will consider a convex mirror and image formation. This is analagous to a concave (diverging) lens.

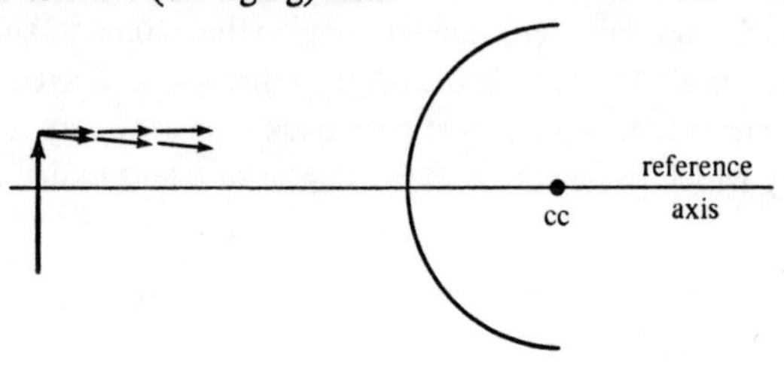

Continue the two rays shown and indicate the reflected ray as they strike the mirror.

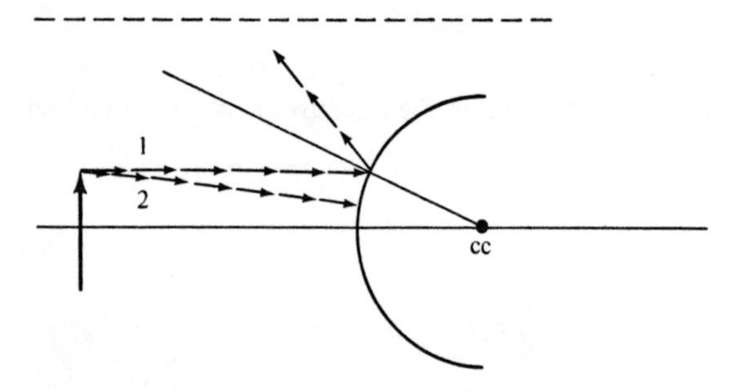

Notice in particular that ray 2 extended in imagination would pass through the center of curvature and therefore just reflect back along the incident ray.

25 Ray 1 in the previous answer is parallel to the reference axis of the mirror. Extrapolating the reflected ray through the mirror defines the virtual focal point of the convex mirror.

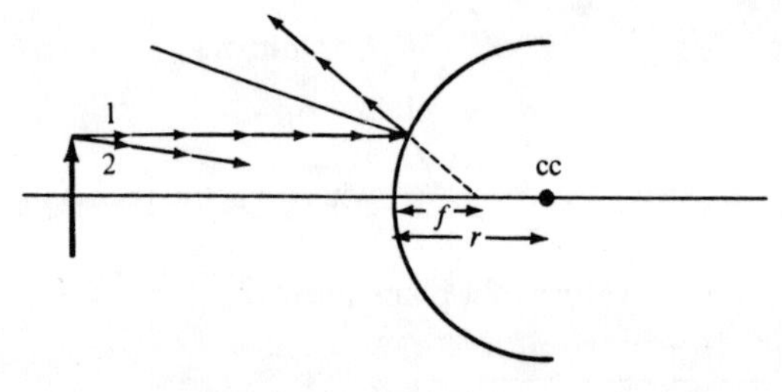

We can see here that $2f = r$ as indicated in frame 21.

Add to the sketch above the extrapolated reflected ray for ray 2.

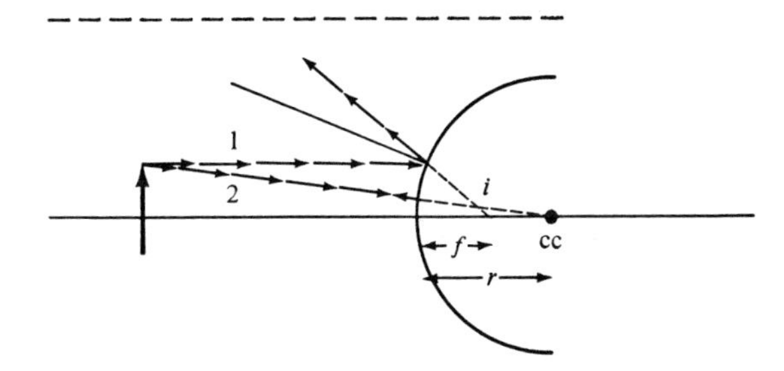

You should *not* show ray 2 "passing" through f. Ray 2 is *not* parallel as an incident ray and therefore does not appear to originate at f. Ray 2 comes in along the perpendicular.

26 Again we consider the light rays which start from some point on an object and try to determine whether these rays do or do not actually converge at some other point in space. Do rays 1 and 2 in the answer to frame 25 converge?

No

27 We say that rays 1 and 2 virtually cross at the point marked i in the answer to frame 25. Thus, these virtual rays appear to come from a virtual image.

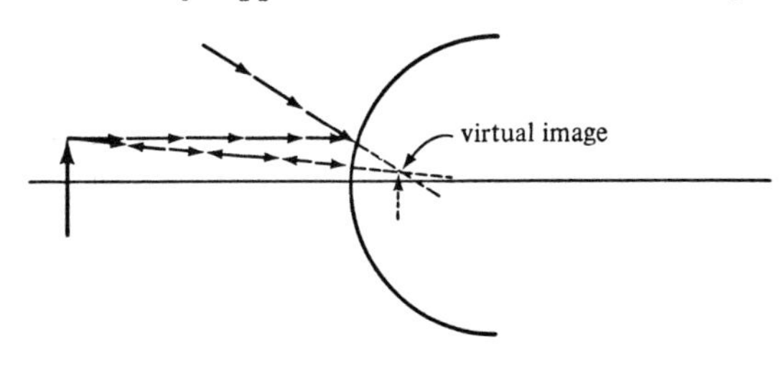

In the case we have been considering, light from the object arrow is reflected by the convex mirror so as to form a virtual image as shown. For example, when you look at yourself in a VW hub cap you see your image as rightside up and smaller than you actually are.

For mirrors we can still relate the position of the object (p) and the image (q) using the relation

$$\frac{1}{p} + \frac{1}{q} = \frac{1}{f}$$

where $2f = r$ for curved mirrors. The sample problems in this chapter will acquaint you with the various sign conventions necessary to solve quantitative problems.

SOULUTIONS TO SAMPLE PROBLEMS

Problem 1

Calculate the critical angle which results in total reflection at the boundary of glass and air.

1.1 To clarify the difference between incident and refracted rays we can consider both possibilities. In the sketch the arrows indicate beam direction.

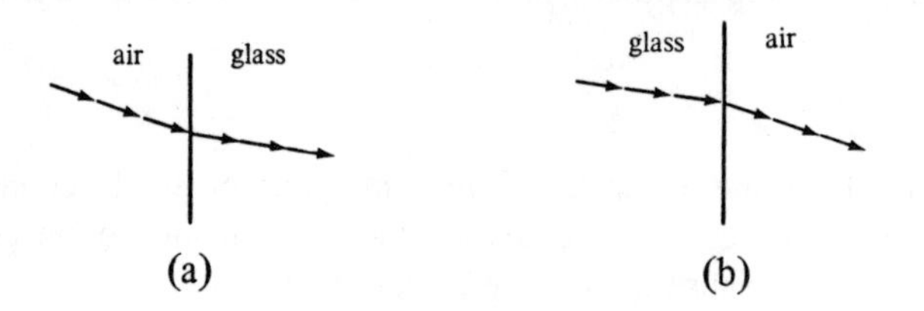

In (a) the incident beam is in ______________ .

In (b) the refracted beam is in ______________ .

(a) air; (b) air (The incident beam is that ray approaching the boundary. The refracted beam is that ray leaving the boundary.)

1.2 When writing Snell's law for refraction

$$n_i \sin \theta_i = n_r \sin \theta_r$$

the terms with subscript i describe the incident angle and the medium in which the incident ray appears. The angles are measured from the ray to a perpendicular constructed at the boundary.

Identify θ_i and θ_r in this sketch.

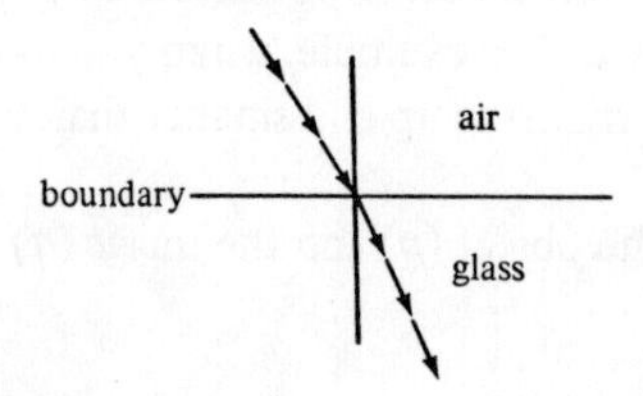

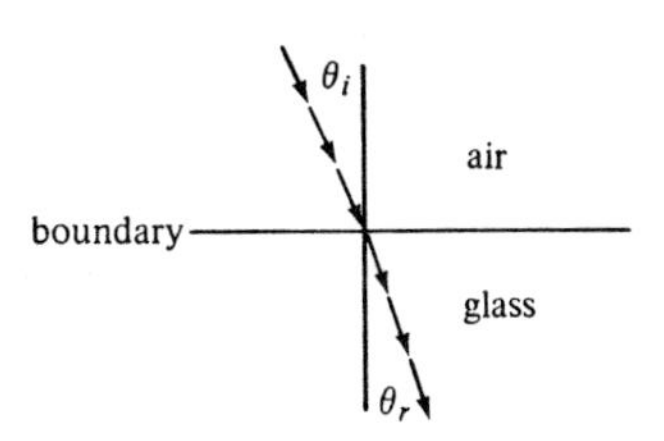

θ_i has to do with incident ray; θ_r has to do with refracted ray.

1.3 Draw perpendiculars at the boundaries located with dots.

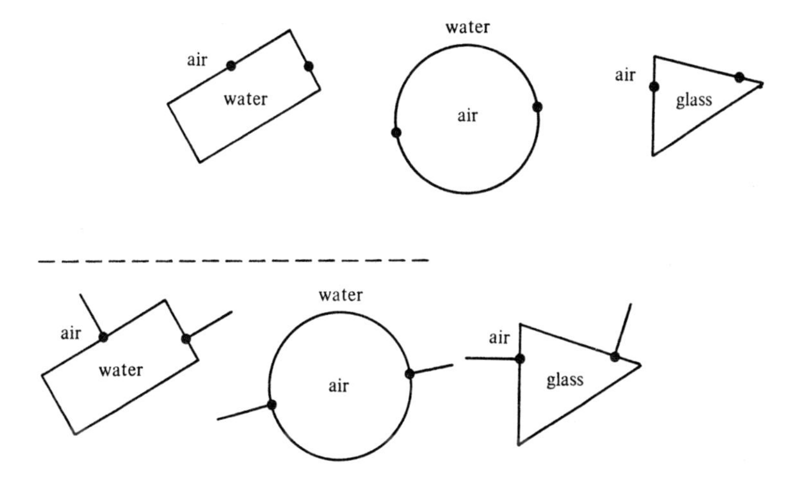

In the case of the circle the perpendicular is an extension of the radius.

1.4 Identify θ_i and θ_r.

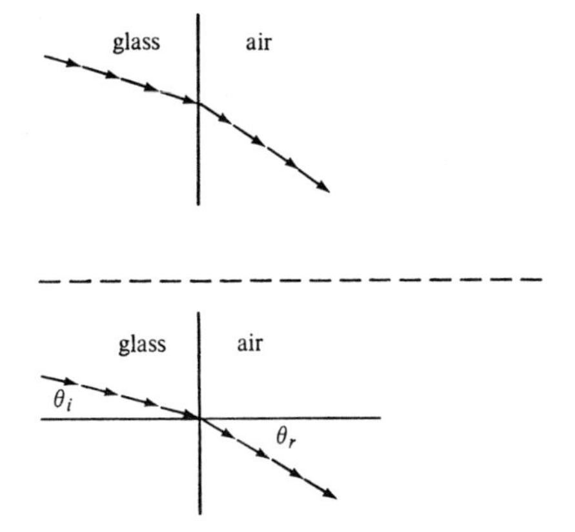

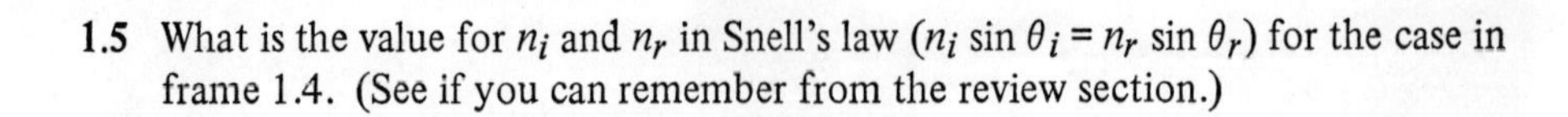

1.5 What is the value for n_i and n_r in Snell's law ($n_i \sin \theta_i = n_r \sin \theta_r$) for the case in frame 1.4. (See if you can remember from the review section.)

$n_i = 1.5$ (the index of refraction of glass which is the incident of medium)
$n_r = 1.0$ (the index of refraction of air which is the refractive medium)

1.6 Rewriting Snell's law we have

$$\frac{n_i}{n_r} = \frac{1.5}{1} = \frac{\sin \theta_r}{\sin \theta_i}$$

Which angle must be larger according to this equation?

θ_r (Sin θ goes from 0 to 1 as θ goes from 0 to 90°. For $\sin \theta_r / \sin \theta_i$ to be equal to 1.5, $\sin \theta_r$ must be larger than $\sin \theta_i$.)

1.7 You will note in the diagram of frame 1.4 that we already included this result in the sketch by showing θ_r larger than θ_i.

From the sketch of frame 1.4 what is the maximum possible value of $\sin \theta_r$?

One (θ_r can be no larger than 90°; $\sin 90° = 1$.)

1.8 To satisfy Snell's law for $\theta_r = 90°$, what is the value for $\sin \theta_i$?

$$\frac{1.5}{1} = \frac{\sin \theta_r}{\sin \theta_i}$$

$$\sin \theta_i = \frac{\sin \theta_r}{1.5} = \frac{1}{1.5} = 0.776$$

1.9 The angle for which $\sin \theta_i = 0.776$ is 42°. Suppose θ_i to be larger than 42°–say 45°. What is $\sin \theta_r$? Use Snell's law again.

$\sin \theta_r = 1.05$

$\sin \theta_r = 1.5 \sin \theta_i$
$\sin \theta_r = (1.5)(0.7)$
$\sin \theta_r = 1.05$

1.10 But there is no refracted angle which can have a sin $\theta_r > 1$. That would mean θ_r is larger than 90°. We see that $\theta_i = 42°$ is the maximum incident angle. Anything larger results in total internal reflection. $\theta = 42°$ is the critical angle for glass. (Note that this result occurs as we go from glass to air.)

Problem 2

Describe fully the images formed by a convex lens of focal length 10 cm when the object is placed first 20 cm away from the lens and then 5 cm away.

2.1 We will begin this problem by discussing certain rays which give a good pictorial understanding of lens behavior.

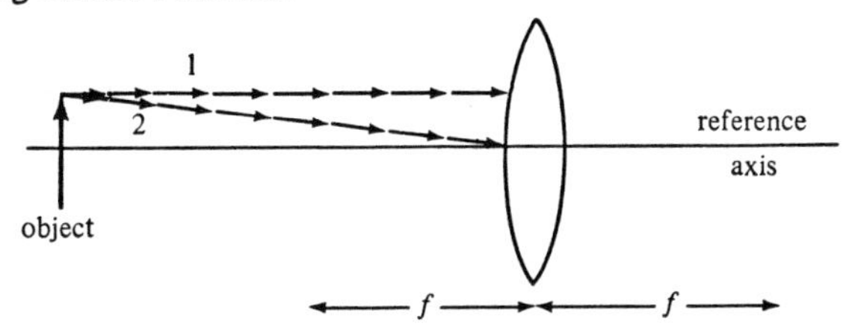

Rays 1 and 2 start at the head of the arrow. Ray 1 is parallel to the axis and ray 2 is aimed at the center of the thin lens. Show the behavior of these two rays as they continue through the lens and pass into air on the other side.

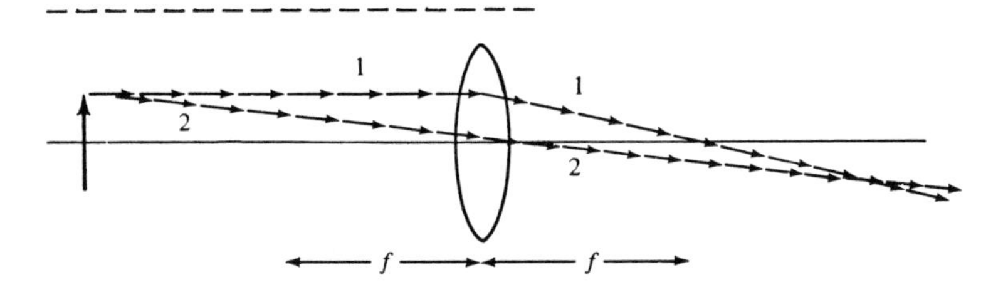

Note that 1 passes through the focal point on the righthand side (as it must since we define f that way). Ray 2 is undeflected as it passes through the center of the lens.

2.2 To clarify the behavior of ray 2 in the previous answer, consider the two extreme cases below.

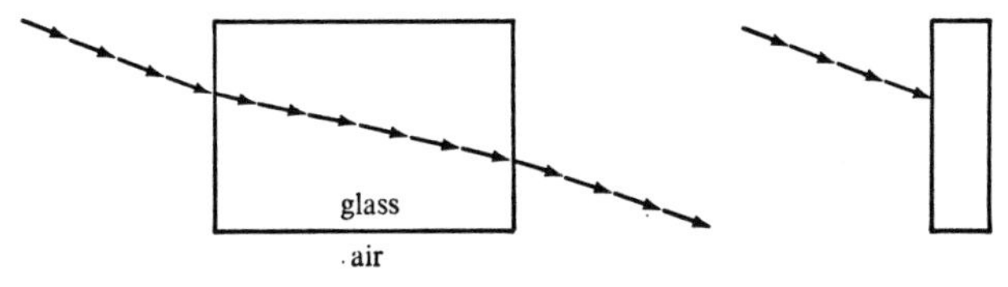

To the left is a thick piece of glass with the path of light ray shown approximately to scale. Complete the diagram to the right where the glass is of the same type only thinner.

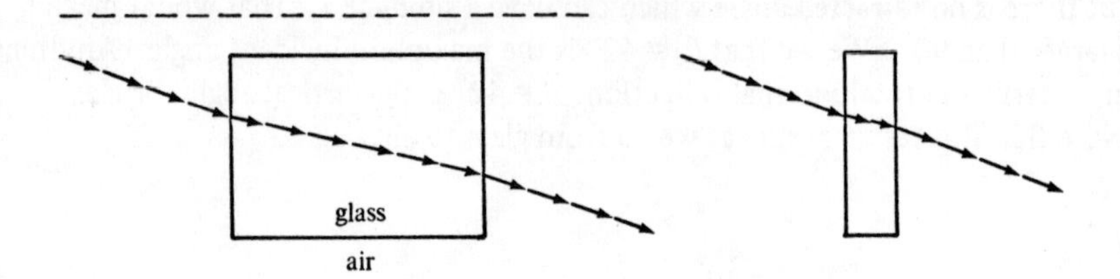

The angles made by the ray in both diagrams are the same. These angles are determined by the index of refraction of glass.

2.3 In the answer above which piece of glass caused less lateral displacement of the beam as it went from air to glass to air? (In other words, which piece of glass had the least effect on the original direction of the light ray?)

The thin piece of glass (Place a straight edge along the rays and compare directions on both sides. For thin glass the ray comes out pretty much as it went in.)

2.4 Refraction at parallel surfaces is small if the surfaces are close together. This is just the situation at the center of a thin lens. Thus, for ray 2 of frame 2.1 there is practically no refraction.

Return to the answer to frame 2.1 and show the path of two rays from the tail of the object arrow.

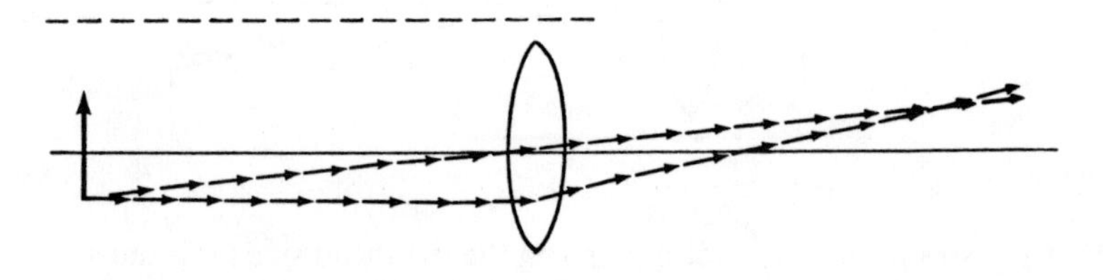

Again we use the same criteria. One ray is parallel and therefore passes through the focal point on the opposite side. One ray passes through the thin portion of the lens (thin as described in the answer to frame 2.3) with parallel sides and is undeflected.

2.5 The previous answer shows that rays from the bottom of the arrows are brought together again on the opposite side of the lens. A screen placed at the point of intersection of these rays would show an image of the arrow bottom. Such an image is called a real image. Is the image inverted or erect?

Inverted

This is characteristic of converging lenses when the object is farther away from the lens than the focal length.

2.6 In this sketch we identify the variables of interest.

p = object distance
q = image distance
f = focal length of lens

The sign conventions are:

p is positive for a real object and negative for a virtual object.
q is positive for a real image and negative for a virtual image.
f is positive for a converging lens and negative for a diverging lens.

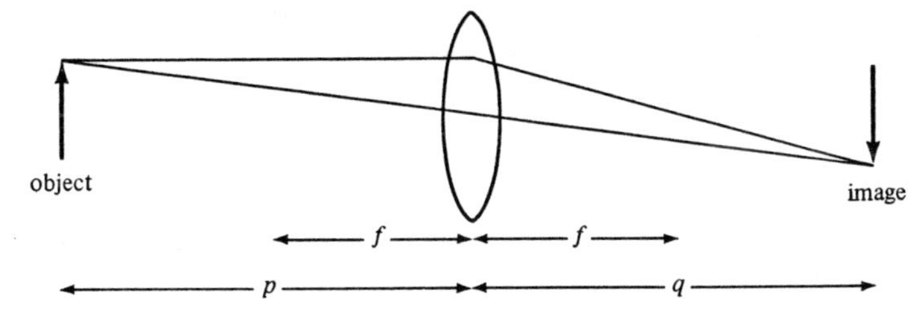

What is the appropriate sign for p, q, and f in the sketch above?

All positive ($+p$ since object is real; $+q$ since image is real–rays come together; $+f$ since lens is a converging one.)

2.7 The object and image relation for converging and diverging lenses is

$$\frac{1}{p} + \frac{1}{q} = \frac{1}{f}$$

where the terms have sign conventions as described in the previous frame.

In part of this problem $f = 10$ cm and $p = 20$ cm. This is the case sketched in frame 2.6. What is q?

$q = 20$ cm to the right of the lens

Note that the ray diagram shows that $p = q$.

$$\frac{1}{p} + \frac{1}{q} = \frac{1}{f}$$

$$\frac{1}{20} + \frac{1}{q} = \frac{1}{10}$$

$$\frac{1}{q} = \frac{2}{20} - \frac{1}{20}$$

$$\frac{1}{q} = \frac{1}{20}$$

$$q = 20$$

2.8 Suppose $p = 50$ cm. What would q then become?

$q = 12.5$ cm to the right of the lens

Since q is positive, a real image would be produced.

$$\frac{1}{50} + \frac{1}{q} = \frac{1}{10}$$

$$\frac{1}{q} = \frac{5}{50} - \frac{1}{50}$$

$$\frac{1}{q} = \frac{4}{50}$$

$$q = 12.5 \text{ cm}$$

2.9 For all values of p greater than f we would obtain q's that are positive (meaning that the image formed would be real). We will delay for a few frames discussing the size of the image.

Now let us draw a ray diagram for the case where the object is real, but inside the focal length.

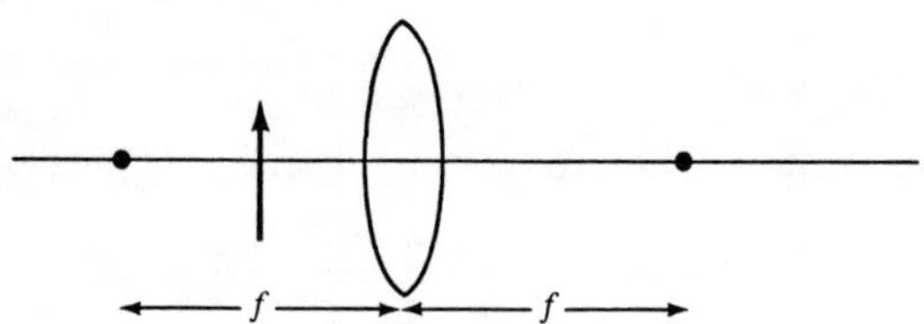

Use the same rays as in frame 2.1 to show what happens as they pass through the lens.

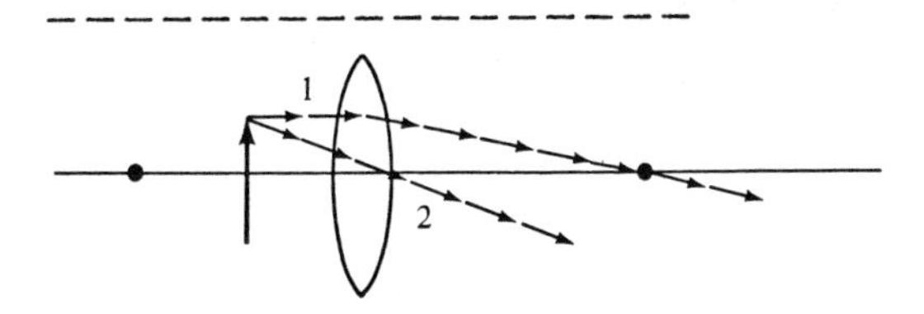

Ray 1 is parallel and must therefore pass through the focal point on the opposite side. Ray 2 passes through center and is undeflected.

2.10 To form a real image, rays 1 and 2 must come together at a point (say on a screen) to form a real image. Will rays 1 and 2 form a real image?

No (They diverge upon leaving the lens.)

2.11 The combination of eye and brain treats rays of light in an interesting way. It seems as though we think of light rays as always traveling in *one* straight line and in particular it is the straight line they follow on entering our eye.

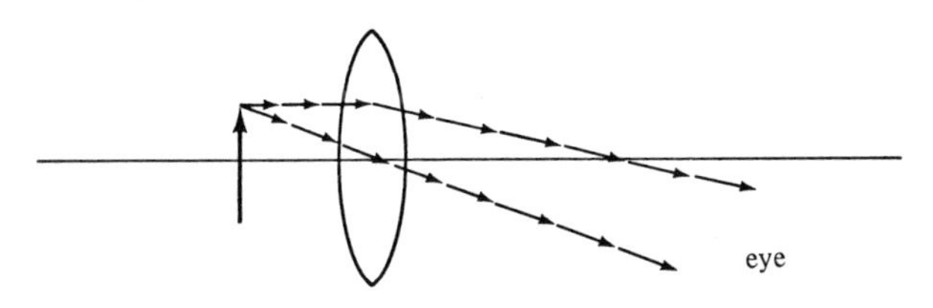

Where does the brain think the rays come from in the sketch above? (The sketch is exaggerated for clarity.)

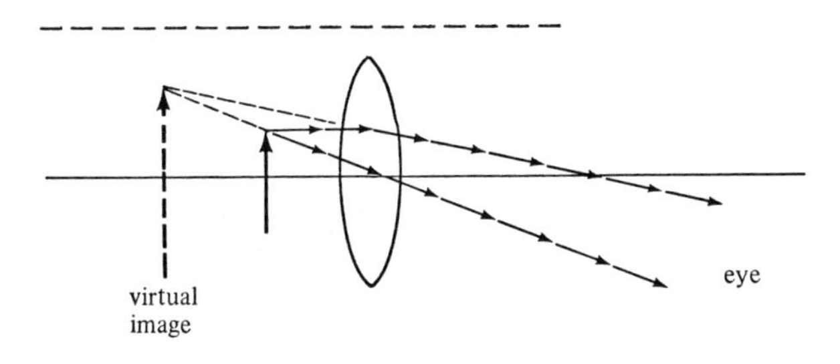

Extending the rays straight back along the path they have to the eye we see that they *virtually* come together on the same side of the lens as the object.

2.12 The idea of virtual images should not be mysterious to you. When you look at a flat mirror you seem to see an image behind the mirror. The ray entering your eye *appears* to come from behind the mirror.

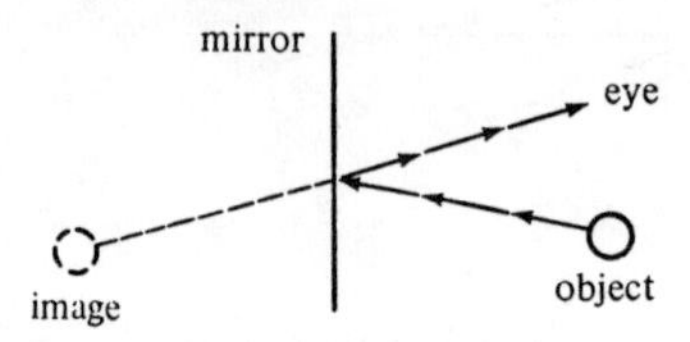

This virtual origin of light rays is called a virtual image. In the answer to frame 2.11, would the image be erect or inverted?

Erect (The rays shown are for the top of the arrow and they appear to originate on the same side of the diagram axis as the real object.)

2.13 Looking again at the conventions in frame 2.6 and calculate q for $p = 5$ cm and $f = 10$ cm.

$q = -10$ cm

The minus sign indicates that the image is virtual and on the same side of the lens as the object. The diagram of the answer to frame 2.11 is accurate. It shows the virtual image at twice p.

$$\frac{1}{5} + \frac{1}{q} = \frac{1}{10} \quad (p \text{ and } q \text{ are still positive.})$$

$$\frac{1}{q} = -\frac{2}{10} + \frac{1}{10}$$

$$\frac{1}{q} = -\frac{1}{10}$$

$$q = -10 \text{ cm}$$

2.14 We can now consider the size of the image whether it is real or virtual. We will do the virtual case.

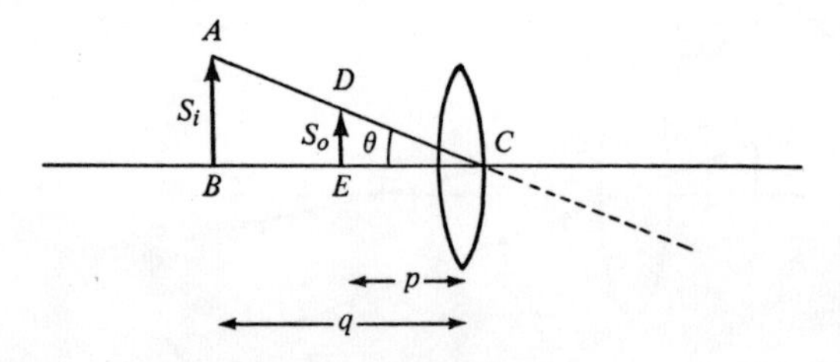

We modify the diagram from the answer to frame 2.11 using only half the object and image and consider only the ray which passes through the center of the lens. Look at the two similar triangles ABC and DEC which have a common angle at C labeled θ. S_i and S_o represent the image and object sizes, respectively.

Write the tan θ using the labels on the diagram. Do so for each triangle.

$$\tan \theta = \frac{S_i}{q} = \frac{S_o}{p}$$

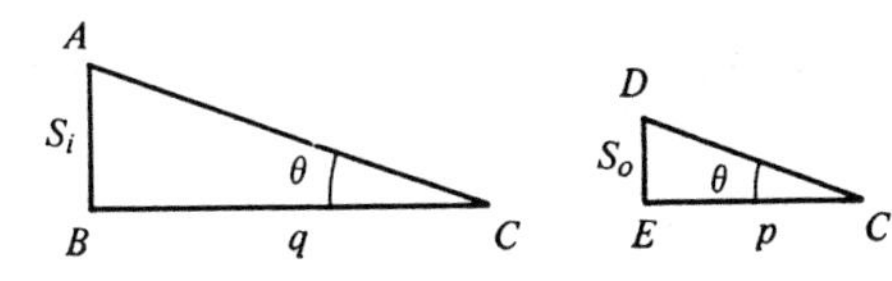

2.15 The answer requires

$$\frac{S_i}{S_o} = \frac{q}{p} \quad \text{or} \quad \frac{\text{image size}}{\text{object size}} = \frac{\text{image distance}}{\text{object distance}}$$

It is usually written

$$m = -\frac{q}{p}$$

where m is the magnification S_i/S_o, and q and p still have sign conventions as in frame 2.6.

(a) For the case $f = 10$ cm, $p = 20$ cm, and $q = 20$ cm, what is m?
(b) For the case $f = 10$ cm, $p = 5$ cm, and $q = -10$ cm, what is m?

(a) $m = -1$; (b) $m = +2$ (A minus m is interpreted as an inverted image. A plus m is interpreted as an erect image.)

2.16 All the characteristics of the images for this problem are indicated in the previous answer. For an object placed outside ($p = +20$ cm) the focal length of a converging lens ($f = +10$ cm), a real image ($q = +20$ cm) is formed on the opposite side from the object with a magnification of 1 and it is inverted ($m = -1$). For an object placed inside ($p = +5$ cm) the focal length of a converging lens ($f = +10$ cm), a virtual image ($q = -10$ cm) is formed on the same side as the object with a magnification of 2 and it is erect ($m = +2$).

Problem 3

A dental mirror of radius of curvature 2 in. is held $\frac{3}{4}$ in. from a tooth. Describe the character, size, and position of the image formed. Should the mirror be plane, concave, or convex?

3.1 Most people have had the experience of a dental mirror being placed in their mouth. What is the purpose of the mirror?

To permit the dentist to view your teeth without looking directly into your mouth *and* to show the dentist an enlarged tooth

3.2 Does the dentist require an erect or inverted image?

Probably erect. He wishes to see the tooth (and his instruments) as they actually are, only larger. He doesn't wish to push on the drill when he should actually pull. (Remember pushing is an inverted pull!)

3.3 For those who are experienced with mirrors, the requirement of an enlarged, erect image implies that the mirror is concave. Plane mirrors provide images which are not enlarged. Convex mirrors provide images which are smaller than the object. You probably don't need convincing about plane mirrors, so let us consider the convex and concave.

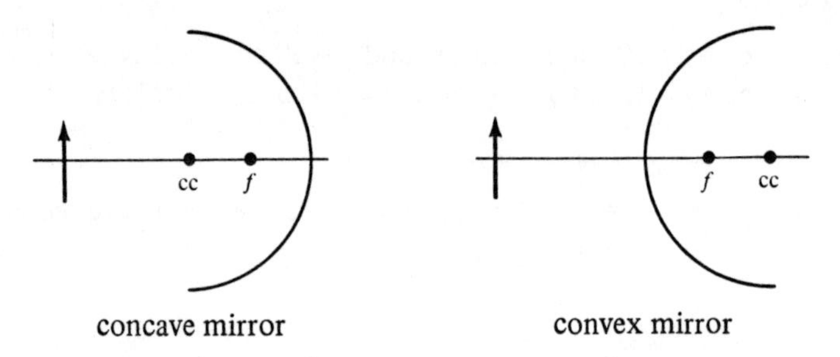

Use the two rays listed below to show whether or not the image formed in each case is virtual, real, inverted, or erect.

(a) A parallel ray incident on each mirror
(b) A ray which is aligned with the center of curvature

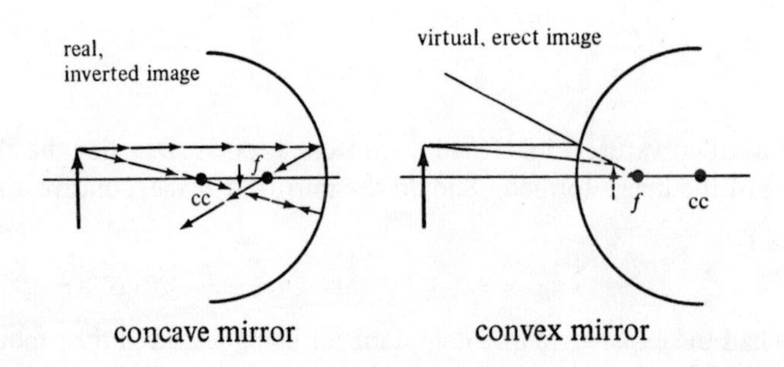

3.4 Would either of the images formed in frame 3.3 satisfy the requirements of the dentist?

No (The inverted image size is smaller than the object size. The erect image is smaller, rather than larger. The dentist wants an image which is enlarged.)

3.5 We can get a hint of what to do if we just think about how the dentist uses the mirror. Is the mirror close to the tooth on which the dentist is working?

Yes (The mouth isn't all that large.)

3.6 The important thing to remember is that the dentist holds the mirror so that the tooth is inside the focal length. In other words p is less than f.

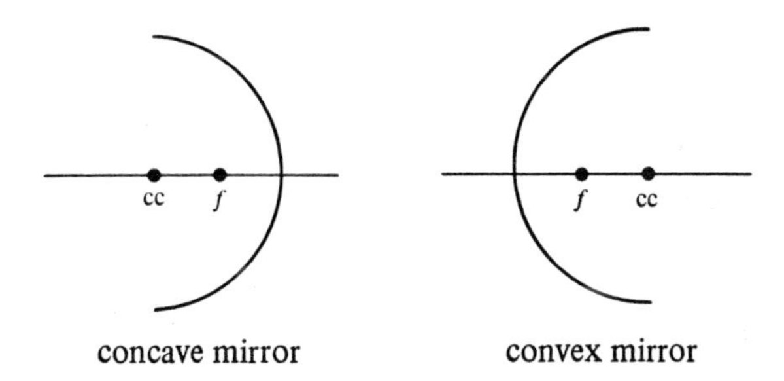

In each case draw an object (arrow) to the left of the mirror surface which meets the consideration described above.

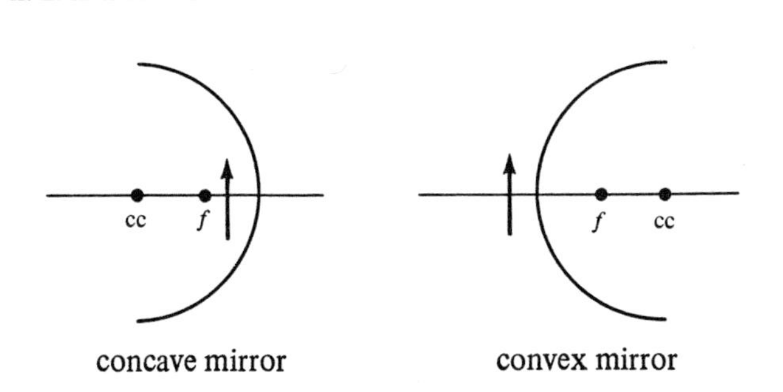

3.7 From the previous answer, show by means of a diagram what kind of image each mirror produces.

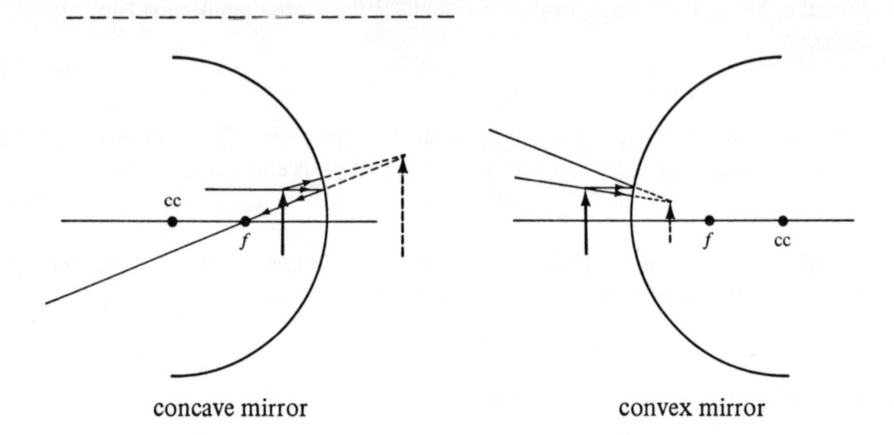

Both produce virtual erect images, but only the concave mirror produces an enlarged image. (Note that we use the two rays described in frame 3.3.)

3.8 A dental mirror is concave and it has a focal length that permits the object, a tooth, to be closer than the focal length. To solve mirror problems we need a set of sign conventions. We have implied by the diagrams thus far that the object (and the light from the object) strikes the mirror from the left.

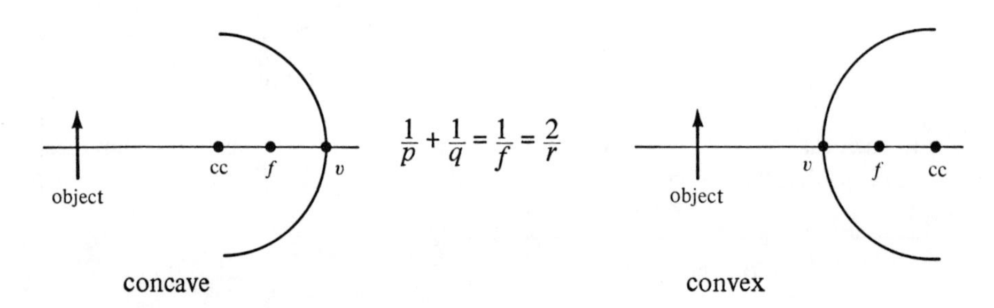

$$\frac{1}{p} + \frac{1}{q} = \frac{1}{f} = \frac{2}{r}$$

concave

(a) $f = \frac{r}{2}$ and is positive.

(b) p is positive and is the distance of object from vertex v.

(c) q is the image distance, positive if real and negative if virtual; measured from v.

convex

(a) $f = \frac{r}{2}$ and is negative.

(b) p is positive and is the distance of object from vertex v.

(c) q is the image distance, positive if real and negative if virtual; measured from v.

Review the answers to frames 3.3 and 3.7 and describe the only mirror and object position which results in a postive q.

Only a concave mirror in which the object is outside the focal point of the mirror

3.9 The main part of the problem is to describe the image formed by a concave mirror with a radius of curvature of 2 in. held $\frac{3}{4}$ in. away from the tooth. Use the conventions of the previous frame to calculate the image distance.

$q = -3$ in.

$$\frac{1}{q} = \frac{2}{r} - \frac{1}{p} \qquad p = +\tfrac{3}{4} \text{ in.}, \quad r = +2 \text{ in.}$$

$$\frac{1}{q} = \frac{2}{2} - \frac{1}{3/4}$$

3.10 Check the answer to frame 3.7 and compare $q = -3$ in. with the appropriate sketch. Where is the image?

Three inches to the right of the mirror (The image is virtual since q is negative and it is enlarged.)

3.11 We determine the magnification in a manner similar to the case with lenses. (See frames 2.14 and 2.15 in sample problem 2.)

$$m = -\frac{q}{p}$$

$m =$ ________

$m = +4$ (The plus sign means the image is erect.)

$$m = -\frac{-3 \text{ in.}}{3/4 \text{ in.}} = +4$$

SELF–TEST

1 A light ray emerges from glass parallel to but displaced a distance d from its incident direction.

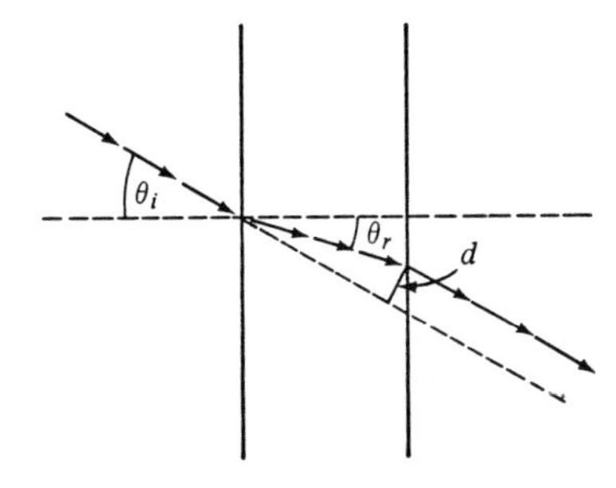

For an incident angle of 60°, calculate d for glass with $n = 1.5$. The thickness of the glass t is 20 cm.

2 Most people have looked in "cosmetic" mirrors which produce a large, erect image. What is the focal length of such a mirror if you appear 50 percent bigger by holding the mirror 30 cm (about a foot) from your face? Show by a diagram how the image is formed and therefore what type of mirror it is.

3 Determine by a carefully scaled diagram the position of the image of an object 8 cm long which is located 30 cm from a convex lens of focal length 10 cm. Show from the diagram that the thin lens formula $1/p + 1/q = 1/f$ is correct and that $m = -q/p$.

Answers to Self-Test

1 $d = 10$ cm

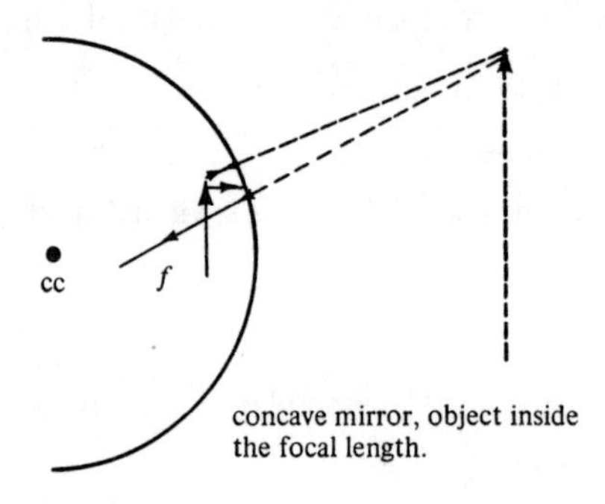

2 $f = 90$ cm

3

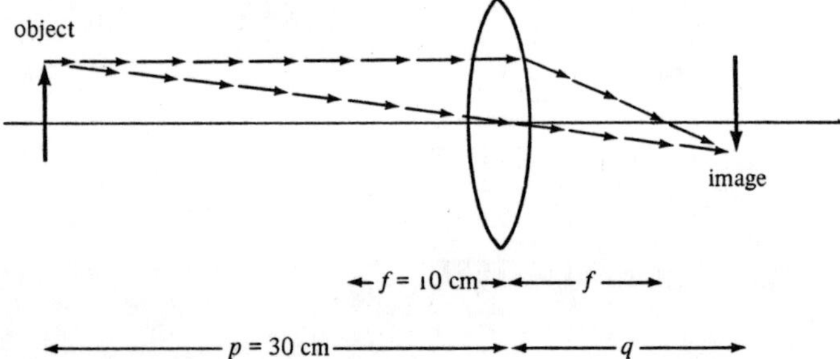

$\leftarrow f = 10 \text{ cm} \rightarrow \leftarrow f \rightarrow$

$\leftarrow p = 30 \text{ cm} \rightarrow \leftarrow q \rightarrow$

From the diagram, $q = 15$ cm

$$m = \frac{\text{size image}}{\text{size object}} = \frac{4 \text{ cm}}{8 \text{ cm}} = \frac{1}{2} \text{ inverted}$$

From the formulas,

$$\frac{1}{f} = \frac{1}{p} + \frac{1}{q}$$

$$\frac{1}{q} = \frac{1}{10} - \frac{1}{30} = \frac{1}{15}$$

$$q = 15 \text{ cm}$$

$$m = -\frac{q}{p} = -\frac{15 \text{ cm}}{30 \text{ cm}} = -\frac{1}{2}$$

Index